# RISK BASED DESIGN FOR
# SAFE DEVELOPMENT
OF RELIABLE AND ENVIRONMENTALLY FRIENDLY INLAND
# WATER
TRANSPORTATION SYSTEM

# Risk Based Design for Safe Development of Reliable and Environmentally Friendly Inland Water Transportation System

Dr. Oladokun S. Olanrewaju, CEng, CMarEng

Co-authors: Ab Saman A. Kader

Rev. date: 10/10/2013

**To order additional copies of this book, contact:**
Xlibris LLC
1-800-455-039
www.Xlibris.com.au
Orders@Xlibris.com.au
504629

# CONTENTS

# 1

# Introduction

## 1.1 Background

Early human civilization use renewable natural resource of soil, waterways, sun, thermal, tide, wind and current for various human needs. Modern technology eventually replaces renewable sources. Renewable natural source requires less energy and produces less waste. Recent system work has seen the trend of activities moving back to the use renewable means in order to sustain the planet and mitigate the challenges posed by population growth, climate change and ozone depletion which are product of the modern technological process.

New source of energy and energy efficient sources include the use of waterways for transportation. About 1,398 cubic million km of water cover three quarter of the earth. The oceans and the seas that connect the lands cover about 71% of the surface of the earth, 94.7 % of which is the oceans. About 5% is the fresh water, where 4% lies in the polar region. Less than 1% makes up all the water in the lake, underground, rivers, moisture in the atmosphere and soil vegetation this makes river resources management including water transportation a very important part of human activities (Brown et al, 1996).

Harnessing renewable natural resources for human need become important at a rate difficult to exaggerate. Safety, environmental risk and reliability based design is required for sustainable system design. System integration, recycling, material, energy and other requirement to satisfy efficiency are also important to be considered as integral part of design and use of renewable natural resources like Inland Water Transportation (IWT). Recent research reports revealed that the potential of IWT usage will increase due to various benefits related to safety and environmental factors. However, implementation of IWT requires risk and reliability analysis where and mitigation which can be achieved through outcome of model in this study (WRI, 2001).

## 1.2 Research Need and Gaps Analysis

### 1.2.1 Current Situation

The ocean and rivers provide human livelihood with source of freshening winds and current. The rivers that provide fresh water are far more vulnerable to polluting activities that have run off into them too much harm that can feed bacteria and algae and which exhaust the oxygen. The ocean may cease to serve its purpose if care is not taken to prevent and control pollution. Marine pollution can be prevented through waste processing or prevention of accident that leads to potential release of affluence in waterways.

Contemporary time has seen alarming environmental revolt that is presently calling for attention and concern about the biosphere world, a condition that call for need to use advantage of human improved knowledge and civilization to develop proactive, reliability and predictive based system for sustainable design. Such system should be able to limit uncertainty components of system design. Issue of environmental problem is linked to water, air and soil. Environmental problem touched every part of human activities including operations, construction, development and design of new marine systems as well as modification of existing system. Marine environmental problem can happen in two distinct ways (GESAMP; 1997). Accidental that includes collision and pollution (consequence of accident). And operational which is the function of point form release of pollutant; oil pollution, Green House Gases (GHG) emission.

Ships and shipping remains a very important instrument for mobility on water. Problem in shipping could lead to largest spill of catastrophe in human history. If ships could no longer transit the waterways, there will be shortages of power, heat and food in days and weeks. Recent years have witness economic of large scale of ship coming from improved trade. The significance of these trends is that more and larger ships will continue to use waterways of the world for the foreseeable future. But there are also limits on size of ship that a channel or inland water can accommodate. Therefore, means of determining when special measures must be imposed on handling ships in order to ensure the safe, efficient, and environmentally friendly use of the channel is a necessity. One way, is employment of risk and reliability based design model that give leverage for high know to unknown ratio in waterways design and planning activities (Kader, 1997).

Environmental issues that result from condition of not safe become really serious because of realization that environment has no border while safety cost money. Other concern is associated uncertainties, point form pollution and negligence to that could lead to accident consequential pollution. The occurrence of accident results to cumulative impact that reciprocally results to degradation, climate disorder, sea level rise as well as threat to natural environment, the survival of the planet earth and the future generation (Guedes, 2001). Maritime industry has been well regulated through previous global coordination work of International Maritime Organization (IMO) and reciprocating local systematic port and flag state implementation. Inland water development management could benefit from the work done in international regulation for governing use international water transportation. This includes vessel operation couple with various measures being promoted by Permanent International Navigation Association Congresses (PIANC) for coastline activities (PIANC, 2007).

## 1.2.2 Normative Situation

A new age of knowledge and sensitivity about need for system reliability and environmental safeguard shows that the reality of the chemistry of matter in pollutant and associated reaction is full of unknown. For example the knowledge of where, what happen somewhere will end up at or to is very limited. Past IMO work in relation to Safety of Life at Sea (SOLAS), Marine Pollution (MARPOL) and Standards Training Certification and Watch Keeping (STCW), International Labor Organization has focused on concept that pollution is about accident and accident is about pollution, because the later is the cause of the former.

New established annex to MARPOL arise as a result of pressure of consequence of accident, global warming, transfer of indigenous organism and ecologic impact. Environmental conservation difficulties posed by variability of geographical environmental factors, unavailability of data, cost and knowledge of chemical reaction leading to degradation are also a matter of concern. This makes hazard quantification that can give system definition, objective and reliability to formulate necessary rule that will make the society function better in sustainable manner a necessity (Pittock and Wratt, 2001).

Previous human development and system design activities has been much more about conventional reactive process that is based on incident and accident. This reactive and try and error process is full of oblivion to the reality of sharing and sensitivity about things that support life has that consequentially loaded systems with risk. For example, recent economic of large scale due to demand for ships and other technological pressure has been mostly unilaterally addressed. While risk related to consequence and frequency of accident is rarely considered. This can be facilitated though use of scientific system based analysis method. Demand Pressure Impact Response Action (DPIRA) is part of system factors that naturally streamlines or relaxed regulation muscle (Guedes, 2001). Ships that resulted from certain demand like the case large tanker development demand in the 1990s, rarely take issues relating to channel condition, subsystem level analysis requirement, system interaction with environment, integrated system components, rate of failure and consequence of failure into consideration (Guedes, 2001).

Contemporary time demand that proactive based philosophy under safety and environmental framework should be exercise on all level of system life cycle, including design, construction, operation and disposal. Selection of all element of the life cycle should be responsibly done to meet reliability and sustainability requirement. Material properties and environmental impact should be addressed to reduce propagation of degradation. The proactive process also required that pollution impact of ship end of life scrap product to the environment and community should be mitigated (IMO, 2009). Channel design criteria, ships controllability in dredged channels and ship maneuverability to minimize collision should be considered in iterative spiral design process. Energy and ship with regards to fuel should be responsibly selected to compliment land base issue of environment solution to cut GHG and ozone depletion, as this is much linked to ships fuel, machineries and propulsion efficiency (IMO, 2009). Introducing IWT is one unique system solution to confront the above challenges of safety and environment.

### 1.2.3 Expected and Desired Situation

Previous work on IWT studies focus more on comparative cost and technical advantage. Less study are made on environmental advantage of IWT. Recent study (Eastman report) outcome on IWT revealed that consideration of environmental advantage for development of IWT ends up with less cost compared to other mode of transportation. The report is not implemented due to political motive behind various government policies. The philosophy of sustainability make environmental issues including the use of inland water transportation, to become important part of today's system design, operation, policy and management (Eastman, 1980). Since, the Rio declaration, the themes on need to improve system based on equity of regulation, social economic equity and management of environment and sustainability. This is in turn changing technological ways of doing things in all walk of human endeavor including management of water resources under IWT. Public awareness about environmental consequences of several types of activities has equally increased. This has remained a driving force for decision on new alternative technology including campaign to shift transportation mode from other mode of transportation to IWT (Kader et al., 2004).

Associated with environmental impact examination are problems of uncertainty, accumulation of point form degradation, high risk, disputed values, and short time line for decision, geography and human reliability. This make the use of constants in risk work a careful matter. This also calls for dire need of new philosophy, science and geographical consideration to design, maintain and prevent alarming system failure that could lead to inconvenient environmental consequences. Development of new waterways requires reliability based evaluation that deals with the risks associated with ships that are plying them. It require securing information sharing avenue that could help development policy of recommendations that could address the way channels are laid out, enlarged and deepened. And how ships of various types using them should be designed and handled.

The study focus on the development of safety and environmental based multi criteria risk assessment, that include use of hybrid technique from deterministic, probabilistic, stochastic based analysis leading to goal oriented comprehensive models for reliable, efficient and sustainable IWT. The model consider factors that deal with safety and prevention of environmental problem that followed accident in relation to navigation channel, vessels and other use of water resources linked to IWT for sustainability, since IWTS cannot stand a alone.

Risk and reliability model focus on life cycle elements that can optimise design, existing practice, incorporate innovative entity that can further facilitate decision for policy accommodation for shift in transportation mode and implementation of IWTS. Sustainable system for IWT requires equity between eco-technology, eco-development, environment, eco-energy, and economy and community involvement. The model outcome can help in developing safety and environmental management that support future guideline for reliable and sustainable IWT design and planning (Laurel et al., 2006).

Human sustainability and survivability need obligation under demand, pressure, impact, response action (DPIRA) to address developmental equity issues including efficient use of natural resources, community participation. It is imperative to employ predictive and reliability based prevention, protection and control measures for complex and dynamic system design like IWT. For new systems, design effort focus on use of all means of preventive measure through use of proactive risk based method to prevent occurrence of accident, and use of control measure to deal with remnant risk.

Proper planning for maintenance, disinfection processing that render discharge acceptable should be arranged under emergency response to various discharges from accidental scenarios. Design drive towards avoidance of loss of propulsion, maintaining ship maneuverability at good efficiency and safety in port, because the consequence of low probability accident of complex nature of IWT system could lead to environmental catastrophe of unbearable casualty and environmental degradation. Risk based design is one of the issue under discussion in international convention regarding optimal and sustainable system design analysis. This is one reason the study is important. In ship and navigation channel design work, 80 percent of a product or service total system life cycle cost is set at concept design stage. Incorporating sensible protective, preventive and control measures in channel, ship design can optimize previous design methods (IMO, 2004). Reliability and risk work should be incorporated as much as possible to achieve efficient and sustainable IWT design. The study outcome is a risk and aversion and reliability model required for optimum sustainable and efficient IWTS.

## 1.3 Research Problem Statement

IWT is considered the cleanest of all transportation modes that exist today. Viable development of inland waterway and its integrative components has become issue of necessity in today's transportation demand. This is due to critical state of environment which is currently attributed to human activities. Especially, issues related of GHG release. There has been discussion regarding IWT development in the past. For different reasons, they have been shelved due to political elements. However, since this age of awareness and knowledge has revealed that no one is safe under environmental reaction. Also if there is anything that is driving technology of the 21st century much more than speed, miniaturization energy and economy, it is the environment.

This include comparative environmental advantage signature of IWT over other mode of transportation leading to potential rise in use of IWT has equally shown sign of transportation system modal change to the use IWT.

Previous risk assessment for waterway related activities, are checklist and qualitative based to construct risk matrix that are converted to risk level. Risk is defined as product of frequency and consequence, previous quantitative risk analysis model focus on either frequency or consequence, others do not consider cost. This study focus on finding answers to quest of today's safety, reliability and sustainability (energy, economic, ecology and environmental) need that hope to facilitate positive policy that allow modal shift of modal IWT. This study analyse frequency and consequence that represent risk of vessel movement in waterway and deduction of alternative option from risk control analysis. The study considers benefit derived on need to face issue of congestion, energy, economy, climate change, global warming and ozone depletion. In view of the foregoing, the problem being addressed by the present work may be stated as follows; "A further investigation into development of sustainable IWT model using hybrid of risk and goal based safety environmental reliability analysis for IWT is warranted. This requires taking into cognizance quinto bottom line of environment, safety, social and economic as well as life cycle, ageing of complex IWT system factors. The research also requires total system analysis through employment of hybrid techniques from tools like failure mode, fault tree, event tree, stochastic and reliability tools for effective diagnosis, analyses, assessment and evaluation of sustainable system".

Developing IWTS require complex plan that need to deal with issue relating environmental friendly ships, structure, channels, information system, integrated components, terminals, mooring, canalisation. The research aim to develop safety and environmental reliability qualitative risk assessment and quantitative risk analysis, mitigation, alternative hybrid option selection for sustainable environmental sound eco IWTS design. The model will also consider marine vehicles, the waterways, human and technological elements of design, operation, maintenance and construction.

Assessment made through goal based prediction and investigation hope to help in the design of new efficient sustainable IWT. The model could provide solution to contemporary call for long term sustainable, environmental friendly legislation requirement for implementation of IWT modal shift from other mode of transportation. The model could also provide subsequently preparation of the marine industry and the nation for compliance in an age of sustainability, knowledge and globalization.

## 1.4 Objectives of Study

The main objectives of this study can be phrased as follows:

i. To study, pressure, response and action associated with IWT development and subsequently examine, evaluate current situation hazard and changes required for the development of IWT. This objective will use relevant available data and information that comparative advantage, form and level of degradation studies that identify, assess and evaluate risk of accident occurrence and compare current tradition in failure mode leading to collision in waterways.

ii. To analyse accident frequency and consequence leading to safety and environmental risk preventive model. The analysis consider safety, life cycle requirement, hazard, environmental degradation, regulation and public concern needed for development of IWT that take into account relationship between risk and reliability, dominant parameter require for sustainable development of IWT.

iii. To evaluate risk, cost, benefit, technical, and environmental sustainability based on hazard, degradation or damage, and the role of new innovative eco technological to improve existing methods. This objective consider total risk approach and cost quantification of benefit derived through inductive scientific, reliability and stochastic approach, incorporation of uncertainty, policy requirement and community participation in risk process for maritime safety and environmental protection of IWT.

iv. To deduce risk control option leading to recommendation of viable and feasible initiation of regulatory framework required for implementation of transportation modal shift to IWT and associated development and management. This objective considers requirement for the design change, best practice scientific knowledge, probabilistic prediction t and uncertainty require for cost effective risk control option for the development IWT.

## 1.5 Scope of and Limitation of the Study

The scope of the study is as followed:

i. The research covers the conduct of needs analysis to identify major requirement and classification of IWT. This includes channel, vessel, terminal, and other support systems. Conduct comparative advantage of transportation mode in term of safety, environmental risk life cycle factors and new initiative to develop IWT.

ii. Carry out extensive investigation to identify level, cause, source and impact of risk and verification of the limitations of the existing methods for waterways design.

iii. The research method involves assessment of safety, environmental risk and ageing factor at design level, and consideration for risk relating to operation, construction, maintenance, economic, social, and disposal require to develop reliability based sustainable IWT model. Deduction of risk ranking from analysis and generation of best options towards development of safety, environmental risk and high level reliability objective require for evaluation of the development of efficient and sustainable IWT.

iv. The model evaluation weight prescribe sustainability requirement for reliable and safe IWT is deducted through iterative components of all elements involve in the system. This include quantification of cost benefit (of channel improvement, social, economic, other use water of resources health and ecological and all other integrated water factors and use of reliable technology required for safety and environmental measure required for model prediction, that meet best practice requirement.

v. Application of the model that involves implementation of the model on Langat River to deduce relevant result, and validation work for reliability analysis of the system, this include iteration on channelization, associated benefit, mitigation for environmental impact, monitoring, compensation for uncertainty of future regulation, human element, management, causal factor from other aspect of the waterways risk.

vi. Risk model is expected to be built based on real data, data suppose to be main limitation of the model, and this limitation has been prevented by using real data from Langat River and PIANC and by incorporating probability and deterministic method in the mathematical model. Data related to site, location and the case study are obtained and used to validate the case study.

## 1.6 Significance of the Research

Rampant system failure and problem related to reliability has brought the need to adopt new philosophy based on top down risk, reliability and life cycle model to design, operate and maintain sustainable system. This make reliable and viable transportation development of IWT is necessary. This research is an extension of past studies on cost modeling of IWT and sustainable maintenance of navigation channel. The importance and contribution of this study are as followed:

i. The research will help in decision making to tackle recent environmental problem (global warming, ozone depletion, sediment contamination, biodiversification, noise, air pollution and congestion).

ii. The development of risk, high level goal and reliability based system model for IWT can contribute to deduce possible corrective actions and preventive measures to minimise and avoid design flaws.

iii. The study of predictive reliability based safety, environmental risk and life cycle model will be helpful to custodians of IWT to operate, improve, and manage IWT in safe and environmentally friendly sustainable manner and provide future opportunity to simulate extreme IWT operation.

iv. The study will help reduce level of unknown and encouraging community participation components of sustainable system design.

v. The novel technique could lead to reduction in workload, cost and time saving, through process of reengineering that can be used to better serve the eco IWT need.

vi. The research hopes to help bring life to IWT and sub sequential achievement of the following advantage of IWT to humanity, economical freight haulage and handling, eco friendly low air emission and low noise pollution, reasonably efficient transport system, low maintenance after initial building.

## 1.7 Research Approach

### 1.7.1 Conceptual Frame Work

The conceptual frameworks employed in this research are IMO`s Formal Safety Assessment (FSA) a derivative from offshore Quantitative Risk Analysis (QRA). United Kingdom Health Safety environment framework (UK HSE) and Goal Based Standards (GBS), are recognized as rational and systematic process for accessing the risk related to maritime safety, protection of the marine environment and evaluating the costs and benefits of alternative options for reducing the risks.

IMO adopted FSA to facilitate transparent and science based decision-making process that provides a clear justification for regulatory measures and comparison of mitigation or measures to be adopted. The main framework for risk based design in maritime industry is the formal safety assessment and goal based design. The main requirement for this approach to modern maritime safety design, operation, maintainability and regulation are:

i. Proactive culture that predicts hazards, rather than waiting for accidents to happen, has occurrence come with cost in for of money and safety to human life, environment and damage to property.
ii. Systematic technique using a formal and structured process.
iii. Transparent that involve clarity and justification of the safety level that is achieved.
iv. Cost effective analysis that involve finding the balance between safety (risk reduction) and the cost to the stakeholders of the proposed risk control options.

The rational for adopting this concept is based on recommendation by IMO that the framework should be consistent in order for the studies to support decision-making process. IMO members have considered FSA to be pre-requisite state-of-the-art conceptual method to assess maritime risk and formulate safety policy to any significant change to maritime safety regulations and it is currently the (IMO, 2002). The risk framework involves evaluation of the risk for human life, on board and in the surroundings of the ship. The framework is implemented using design principles and criteria that evaluate; damage due to prolonged exposures to unhealthy substances or actions, damage injuries and loss of life due to accidents during construction, operation, inspection, maintenance and decommissioning of the ship, damage or loss of tangible items (the ship or part of her, the transported cargo and external objects); and intangible items (e.g. limitations to run activities), The environmental risk, including consideration of pollution due to the normal related to systemic activities developed on the ship during her lifetime; or to accidental events.

The need for proactively has been argued extensively especially since the incident of the Prestige. The conceptual and theoretical framework used in this study are IMO`s FSA and GBS as well as risk analysis UK`s HSE framework. The challenge of risk analysis is to bring the risk to acceptable level and at the same time, derive the maximum benefit which can be quantified into cost (Psaraftis, 2002).

## 1.7.2 Components of the Safety Functional Analysis (SFA) framework

Formal Safety Assessment is considered the prime scientific tool for the development of proactive safety regulation. To achieve the above objectives, IMO's Guidelines on the application of FSA recommend a five-step approach, consisting of:

i.   Hazard Identification
ii.  Risk Assessment
iii. Risk Control Options
iv.  Cost-benefit Assessment
v.   Recommendations for decision making.

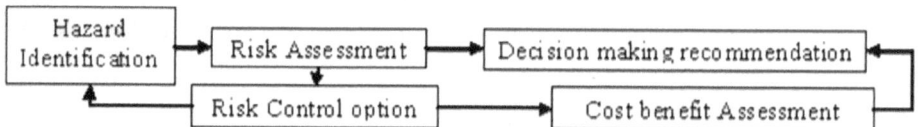

**Figure 1.1:** Components of FSA conceptual risk methodology framework (IMO, 2002)

System functional safety analysis involves assessment of system operational capability, mission requirement and efficiency. FSA involves identification of potential hazard scenarios and major impact to ship, shipping and ship design that could lead to significant safety and operability consequences. It also involves perception on call for policies, chance and procedural major effects. Verification of current design, construction and operations to ensure that risk from identified scenarios meet risk acceptability criteria. If not, to recommend additional system functionality analysis, process, and available technology for control and protection that can reduce risk to suitable level or As Low As Reasonable Possible (ALARP).

FSA targets the following. Elements of FSA according to IACS in MSC are shown in Figure 1.1. Figure 1.2 illustrates components of FSA process. Hazard identification (HAZID) is involve qualitative assessment of the system using PrHA, what if analysis, checklist. Risk assessment is priorization work on identified risk to develop risk matrix. The risk analysis is quantitative modeling of frequency and consequence of system risk. The risk control option is deduction of cost effective decision support. The cost benefit analysis (CBA) is a conduct of cost benefit analysis and lastly is recommendation.

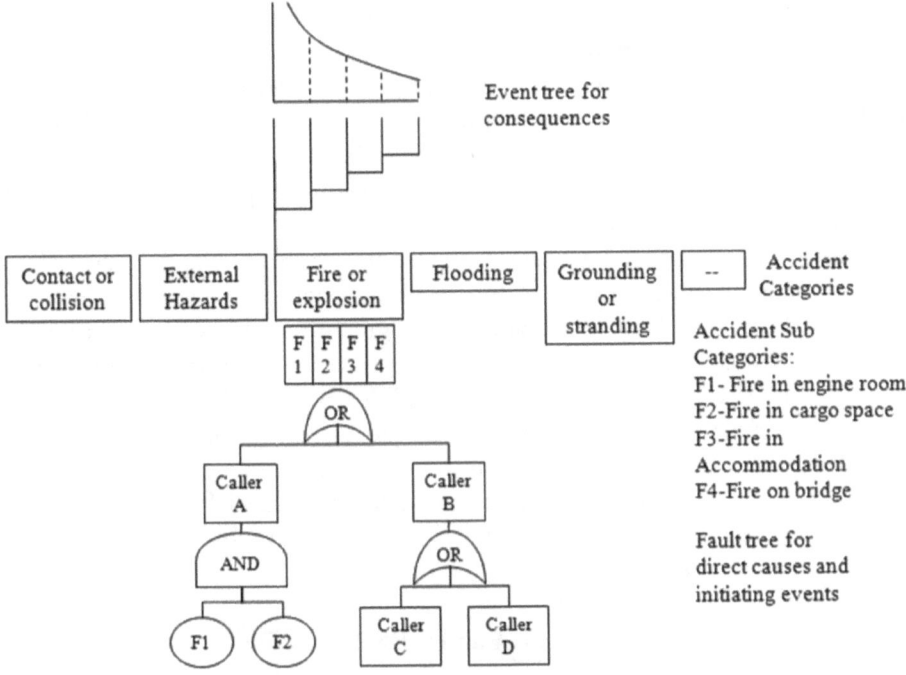

**Figure 1.2:** Risk contributing tree

Figure 1.3 shows high level goal based assessment, it describes the process chart to determine solid objective that cover life cycle, system functionality and ageing factors elements. Regulatory assessment involves checking of the standards required to me met by the IWT. Life cycle Goal based safety standards assessment (SSA) represent the following tiers presented in Figure 1.3.

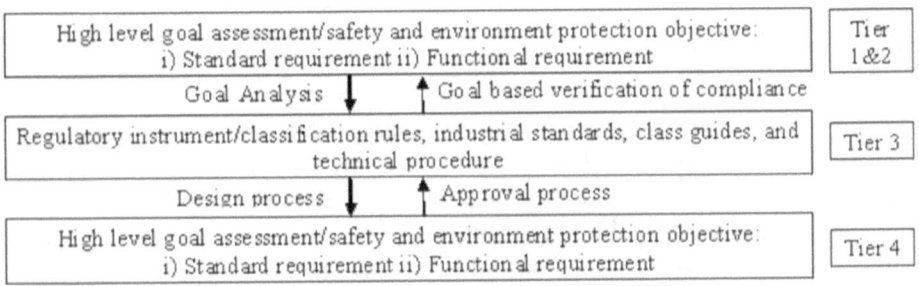

**Figure 1.3:** High level objective safety standard assessment framework (IMO, 2002)

The stage description is as followed; stage I consists of goals expressed in terms of safety objectives. It is defined by risk level. Stage II consists of requirements for ship features and capabilities. It is defined by identified risk level that assures achievement of ship's safety objectives. Stage III is Tier IV and V required to verified system compliance with Tier II. Stage IV consists of primary rules, guidelines, technical procedures, programs, and other regulations for system design and system operation needs. It also involves fulfillment which satisfies system feature and capability requirements. Last stage is Stage V that consists of the secondary code of practice, safety and quality systems that are to be applied to guarantee the specified rules by quality level. In order to select between alternative technical or regulatory solutions to specific problems, the GBS, FSA, the first three (HAZID, risk assessment, risk analysis can fit into the development of high level goals (Tier I) and functional requirements (Tier II). Equally, the (RCOs, CBA, sustainability and recommendations) could fit into Level IV and V to meet the decision require for the system. Table 1.1 shows hybrid matching of formal safety and goal base system.

**Table 1.1:** FSA and GBS connection

| Stage | Goal based standards | Formal safety assessment |
|-------|---------------------|--------------------------|
| Tier I | Overall goal for safety level criteria | |
| Tier II | Functional requirement | Identification of hazard leading to functional requirements |
| | | Risk analysis leading to balanced risk acceptance criteria |
| Tier III | Verification for approval | Risk control option |
| | | Cost benefit analysis |
| Tier IV | International regulation rules and standards | Recommendation for decision making |
| Tier V | Local standards and procedures | |

FSA steps are analyzed by simultaneous description of regulations that goes into standards definition for the system. Both regulatory and functionality requirements will be matched to realize the gap in the system. That leads to setting up high goal objective to determine all risk associated with the system and responsible ways to avoid them. System definition comes with safety, environmental assessment, data collection and procedure. This involves gathering information about the river profile, the tributary, activities and vessel particulars. Components of HSE risk analysis conceptual framework is shown in Figure 1.4, output from high level goal based objectives and hazard identify become main input to the risk analysis.

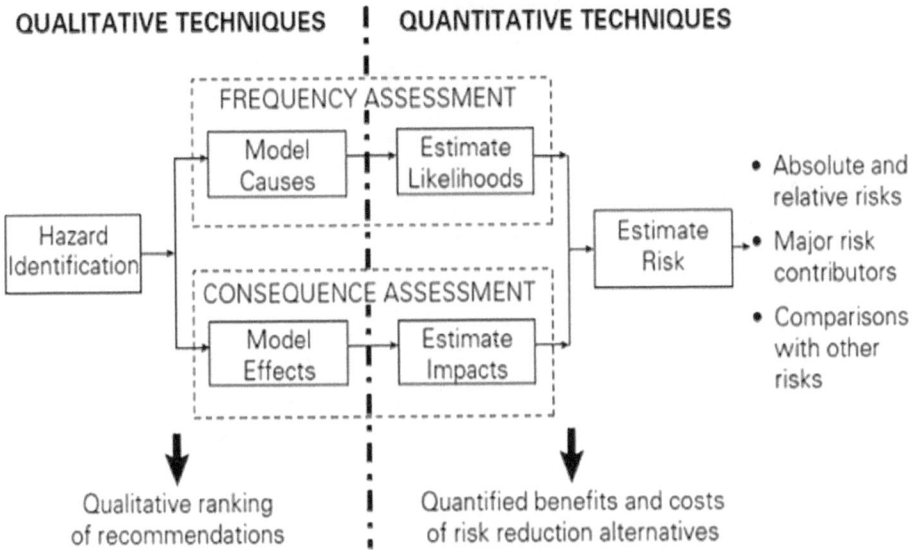

**Figure 1.4**: Risk analysis framework (QRA, 1996)

Risk is defined as product of probability of event occurrence and its consequence.

Risk (R) = Probability (P) X Consequence $\qquad$ (C) Eq. 1.1

Damage from an accident leads consequence to the inland waterway system, infrastructure, vessel, to the environment. The cost of consequences to property, material, environment, related industries, and public perceptions remain the driving forces of development of standards for design against accident and consequence analysis introduced in this research. This includes accident scenarios, process of evaluating consequences, and risk based design acceptability criteria. Figure 1.5 shows ALARP risk acceptability conceptual framework.

Acceptability is judged by examining if the consequences for accident scenarios fall within a specified acceptance range using recommended evaluation and acceptability criteria. The maritime industry acceptability criteria were adopted from UK HSE acceptability (HSE, 2001).

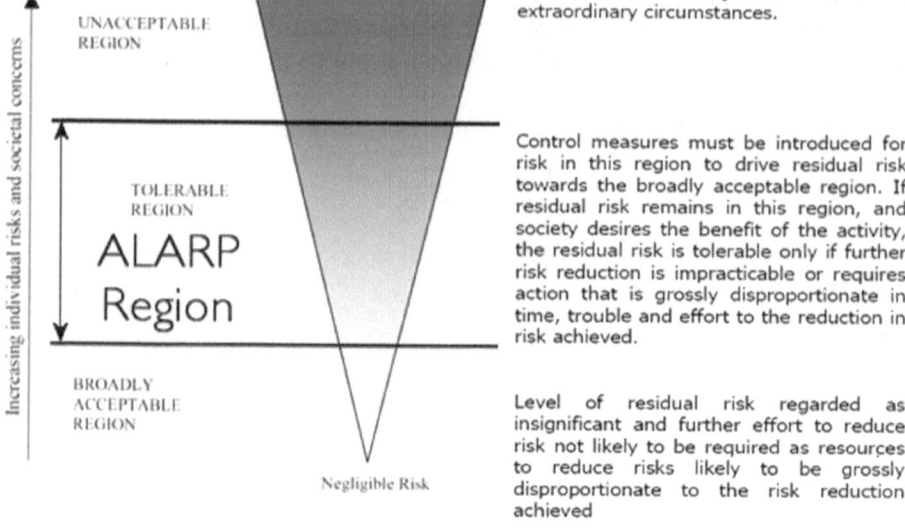

**Figure 1.5**: Risk tolerability framework (HSE, 2001)

### 1.7.3 Data Collection

Data collection for the case involves documents, interviews, literature, master plan related to project documents. Data are also collected through expert consultation and round table discussion with institution like LUAS, MARDEP, and guidelines published by Marine Environmental Protection Committee (MEPC), Marine Safety Committee (MSC), United Nation Environmental Protection (UNEP), Global Environment Studies and Monitoring (GESAM), Environmental Port State Control (PSC), International Safety Management (ISM), SOLAS, International Association of Navigation Congress (PIANC), United State Army Corp of Engineer (USACE), Ministry of Transportation (MOT), Department of Environment (DOE), other related ministry and agencies. Studies publications are also relevant resources used for this research.

### 1.7.4 Case Study

The model developed is tested considering the case study of Langat River to to validate the model by comparing and determining the collision risk aversion feasibility for efficient, safe, reliable and sustainable use of Langat River for transportation. Lembaga Urus Air Selangor (LUAS), management of Langat River indicated that their main problem in Langat

River waterways is collision. Data source from the case study area include safety and environmental parameters relating to vessel, channel and risk associated with development of IWT. The research hopes that the use of Langat River to validate the model will hopefully initiate policy change for transportation shift and integration towards freight IWTS using the Langat River. The Langat River basin is unique river of outstanding length of 200km, located at southern part of Klang Valley. The basin uppermost stretched to total catchment area of 2,400 sq km. Langat River is considered the most urbanised river basin in Malaysia. The basin is divided into ten strategic zones whose population reflects on the scale of economic and social interests. This gives Langat River, the environmental and the population unique potential for development.

The data collected in respect to Langat River to test the model include safety and environmental risk parameters as well as comparative advantage factors useful for transportation modal shift for the design of IWT for freight transportation. The case also consider allowance for future use of the waterways for passenger, water sport operation, advantage accumulated from reduced noise and air pollution, energy use and increase in transportation volume for the region (DOE, 2000). The case can help the logistics of Port Klang and Straits of Melaka riparian water, the rural development and reduction of GHG emission in the region. The study hopes that testing of the model on Langat River will complement previous study made to bring the use of river resource alive. The model developed in this study is a generic model based on PIANC criteria, which is useful for all waterways. Langat River is chosen for case study validation for the following reason:

i.   IWT development is a very complex. Complex systems are required to be developed based on risk and reliability. Testing and comparative analysis of Langat river on the model to assess risk level

ii.  The concept of sustainable balancing of issues of economic, demand, environment, social and technical safety is always recognized by stakeholders as key contributory element for development of IWT. Safety reliability and cost effectiveness assessment components of the model indicate meet sustainability requirement of the system.

iii. Risk, reliability, ageing factor, sustainability and uncertainty in risk process toward development of the model fall under new system based philosophy being adopted by all field of human endeavor for reliable, sustainable system design, operation and maintenance. Hence, establishment of prove of testing of modest proactive base approach on inland water transportation.

iv. Risk and reliability process starts with description of the system and relationship between system, subsystem, man and environment which goes into functionality definition of the real working system. High goal based objective requires balancing and matching of system functionality to regulatory requirement in order to capture gap in the system and deduce high goal objective to close the gap.

## 1.8 Thesis Organization

This thesis is composed of six chapters. The first chapter provides the introduction which includes the background of the study, the case study, objective, contribution and data collection strategy. The significant of the research and its contribution to the maritime industry and institution as well as the need to design and develop proactive reliable based model for sustainable IWT is discussed Chapter two is associated the literature reviews which discuss the literature work on the importance, advantage, potential of IWT, and literature on proactive measure necessary from best practice. The definition of components of IWT, literature work on development of IWT, energy, issue of global warming, climate change and ozone depleting driving today technology and how they touch IWT are included. Information about present situation and past work on the best practice is presented in this chapter.

Chapter three is the methodology consisting of the details of the best practice approach that can achieve high goal objective set to cover potential failure system gap. The research approach for the development for safety and environmental risk model for IWT is presented. It describes process for risk and reliability modeling for establishment of IWT design. The details of the model development process, model physical, probabilistic, stochastic mathematic representation of IWT system and flowcharts for the modeling process is discuss. Procedure for risk data and processing of the data is presented. The case study requirement, especially data related to river profile, other required baseline and environmental data required for the implementation of the model on a case study in order to deduce necessary intuitive risk and reliability based DSS result for system implementation is presented in this chapter.

Chapter four presents the model results. The result generated from data analysis of the generic model and the case study implementation result and the results are presented in various forms such as tables and graphs followed by evaluations. The background information and waterway profile about the case study is described. Chapter five contains the discussion of the results in Chapter 4. The model has been implemented on Langat

River. The importance of matching system functionality and regulatory for determination of gap and formulation of sound goal based objective that target life cycle and ageing factor to avoid safety and environmental risk is presented to highlight strength and reliability of the model. Issues of environmental problems and the need to apply a risk based model towards a sustainable IWT are discussed to highlight applicability of the model. And lastly the conclusion present and summarize the works done in this research and provide answers to question asked by the objectives and proposed recommendation for implementation, monitoring and further studies on the model and the research.

# 2

# Literature Review

## 2.1 Introduction

Inland water transportation (IWT) either in moving people and freight in a sustainable manner is increasingly becoming important and it is one of the biggest challenges for the 21st Century. In an age for sensitive based reactions to change conventional human activities with by proactive based culture and innovative behavior for design, construction, maintainability and operations of system. This study involve top down reliability safety and environmental risk model for sustainable IWT (Ayyub et al., 2002). There is a large barrier to achieve a sustainable IWT safety and mitigation of environmental impacts and dealing with integrative components of water recourses. However, incorporating total risk systems framework and system engineering tools backed with hazard identification, quantitative risk analysis can facilitate achieving the daring need for sustainability.

The research is an extension of work done by (Kader, 1997) on cost modeling of IWT and recommendation given by Sulaiman (2006) on need to employ risk based method to model in master thesis on sustainable maintenance of navigation channel, such top down system consider all safety and environmental risk information associated with inland waterways, their respective impact and stake holder concern that will provides a reasonably efficient, cost effective means for transporting cargo.

Experienced has shown that nuclear, offshore and airplane industries that deal with almost zero tolerant complex risky system has been able to do well with designs with almost zero tolerant complex risky system has been able to do well with designs backed on predictive risk and reliability methods according IMO (1991). Thus, design from first principle deterministic method has always been helpful incorporating natural laws and principle in the risk process. Since, uncertainty is inheriting part of the system, it is imperative to model it separately. Benefit derived from of IWT can be quantified and translated into the cost benefit of risk process (Trbojevic, 2000).

## 2.2 Definitions and Components of IWT

IWT is defined as; it contains physical elements that include waterways, ports, and intermodal network of railroads, roadways, and pipelines. IWT connect the waterborne portions of the system as required. The physical elements also include the vessels and vehicles that move goods and people within the system. The physical network is supported by a series of systems that facilitate the movement of goods and people, provide access for recreation and to natural resources. The principal components of the IWT and risk can be defined as follows (Van der Leeden Frits et al., 1990):

i.   Waterways: include the navigable waters and associated infrastructure useful for vessel traffic.

ii.  Ports: contain marine transportation facilities where vessels transfer cargo, passengers, it can also be used for recreational access facilities and sometime shipyards.

iii. Intermodal network: are linkages at the land and water boundary that allow the transfer of cargo and passengers between transportation modes. Intermodal connections include pipelines, road, and rail access routes.

iv.  Vessels and vehicles: are the transportation equipment that moves goods and people within the system, it includes oceangoing, coastal, and inland vessels, trains and trucks.

v.   Inland waterway users: are the people who depend on the system for their livelihood and recreational access. Community participation and sensitivity analysis of concerned people regarding their account on system past behavior and their recommendation to support implementation, monitoring and adaptability

vi.  Risk analysis: involve taking all practical hazard consideration, using historical data, probabilistic, deterministic, stochastic analysis that include all factors including public health for sustainable system design.

vii. Mitigation to risk assessment: This involves making permanent changes to minimize effect of a disaster, this include immediacy, Prefer and no option sustainable choice including conduct of iteration between risk, cost technical, social and benefit to reach common ground for balance and compensation that will allow things to run.

viii. Panel of expert: reach out to those who are knowledgeable and experience in the area of concern to cover requirement of design and planning, and uncertainty

ix. Emergency response: provide monitoring, information facilities and make sure necessary information is appropriately transmitted to avoid unwanted risk scenario and associated reversibility

x. Hydro morphology: the physical characteristics of the river structures such as river bottom, river banks, the river's connection with the adjacent landscapes and it's longitudinal as well as habitat continuity.

## 2.3 Nature and Development of IWT

The geography of waterways network is also defined by valuable stages for architectural display and great places for public recreation (TTI, 1997). Waterways management is divided into three stages, unregulated water become supply oriented, it remain so, as long as water is abundant. The demand can be satisfied without modifying hydrological regime, scarcity of water with increased pressure of demand for water and water related services, water management become resources oriented and the basis for multipurpose development, and regulated natural regimes of water limit where acceptable stream flow developments are reached.

Inland navigation offers important opportunities to move cargo on river, estuarine rivers tributaries in an energy efficient manner, lower cost of good transportation, per ton per kilometer compare to other mode of transportation. It remains one of the best options available to mitigate problems associated with global warming, climate change, noise pollution as well as congestion. Capacity building of IWT in environmentally and socially friendly manner should take advantage of nonstructural measures like fleet innovation according to Broils (1967). Comparison of mode of transportation gives IWT a lead in eco transportation. This capability has not been fully taped due to lack of political will. Also due to lack of public concern and focus on the benefits of IWT when decisions are being made concerning a choice of modes of transportation. Malaysia has 7,200 km of waterways, most of them rivers, 3200 km are in Peninsular Malaysia, while 1,500 km are in Sabah and Sarawak has 2,500 km. Malaysia economy derived a lot benefit from transportation of cargo through water. A large numbers of cargo types are being discharged in Port Klang. Distributing this cargo via IWT means substantial growth and development for the nations (MOSTE, 2000).

## 2.3.1 IWT Development Requirement

Previous studies on IWT were based on the environmental, cost and maintenance model. However most of them fall short in design technique that is based on reliability for environmental prevention and reduction of safety risk. Past work like the study did by international association of great lake port rarely discuss on how the benefit offered by IWT can be quantified and translated to cost. Thus, recently, the European and the United States has focused on new direction on innovative design of the risk top down model (European Commission, 2006). But these dedicated approaches are yet to be seen in developing world. In this study the IMO conceptual framework is adopted to develop the model. Availability of convincing environmental and reliability work could give boost to decision support and positive policy trajectory towards use of IWT. Previous methodology for marine system model focused on area of such as Port Sector, Ship building and ship repair sector and IWT Sector (IAGLP, 1972).

Considering the case study in this research, most work done on Langat River under the integrated water resources master plan for use of water resources focused on environmental data collection and flood mitigation. But no study is carried out on risk of navigation or collision avoidance for river transportation. There is a recent study that investigated potential for use of Langat River for navigation. The data from this study shows that nothing is done on risk of navigation, system reliability and life cycle accountability. This study intends to build on and incorporate scientific based risk analysis on previous work done by Kader (1997) on cost model, maintenance of navigation channel and other waterways studies. The best practice methodology to assess and analyse marine environmental safety compliance, quality control requirement for sustainable and reliable system is addressed by using " pressure, state, response" approach (DnV, 2000), where pressure is the demand placed upon the marine environment and its resources by users, state is the current conditions resulting from these pressures and response is what is being done to address the pressures problem.

The ocean and the ports have been managed through IMO, PSC, and FSC in responsible manner. Most IWT has been under utilised and under maintained especially for transportation. The Butts et al (1992) and Eastman (1980) studies outcome promise potential for rise in the use of IWT, this include logistics for short sea and supply vessel operation to accommodate requirement of growing vessel size and expanding deep sea operation, also the use of high speed craft for recreation, and the use of barges for freight transportation. IWT development represents complex and dynamic system that require be designing and establishing based on concrete proactive approach. Table 2.1 shows typical risk components in IWT project phase.

Preliminary safety and environmental risk assessment is best addressed through; investigation of existing situations, regulations, policies in place, survey on stake holder requirement on acceptability and procedures. Also recent call for policies change, major obstacles, effects, impact to ship, shipping, ship design, ongoing commitments, performance, conformance, available technology for control, prevention and protection of environment, collect data, create, organize the data quantitative safety and environmental risk assessment of system and subsystem capability, analyze data quantitatively through probabilistic, stochastic and predictive methods to deduce required mitigation, perform Risk Cost Benefit Analysis (RCBA) to deduce mitigation and risk control, and provide answer to current IWT safety and environmental problem. Table 2.1 shows components of risk analysis

Towards sustainable reliability system design, it is preferable to use hybrid of deterministic, stochastic and probabilistic methods that could help improve the existing methodology. Such method create positive environment that will cover most uncertainty required to compliment historical data and give a more representation of risk factor in system s risk matrix. Hybridization of models and beyond compliance approach is also a plus for the best solution of sustainable maintenance of navigation channel under monitoring of designed system. It involve the employment mixture of method comprising goal based objective, safety assessment and qualitative and qualitative risk analysis through use of tools of deterministic, stochastic, probabilistic, statistical, historical, expert input and community participation in safety and environmental risk and reliability based design models (Zamani, 1999).

**Table 2.1:** Components of risk analysis

| Components | Purpose/Process |
| --- | --- |
| Risk analysis | Involve the overall process of risk assessment and risk management, including screening and scoping |
| Risk assessment | Involve qualitative process of identifying risk potential to quantitative risk characterization |

| Risk screening | Involve the specification and setup of a general framework for managing risk |
|---|---|
| Risk evaluation | Involve: Scientific evaluation of risks through use of stochastic process Public/political evaluation of risks |
| Risk management | Involve: Process of identifying and selecting measures Procedures for implementation and evaluation of measures |

Recent time has seen environmental calamity and abnormal environmental behavior that leads to the consensus of scientist has agreed to link such events to human activities. The Planet is made up of the biosphere and the techno sphere. Human inherited the biosphere that support human to live. However, there is benign for care needed for the biosphere due to lack of knowledge and sensitivity about relevant interaction. Human created techno sphere whose by product are responsible for the effect of climate change and consequential planetary natural accident events.

## 2.4 Threat of Climate Change

GHG impact is expected to cause adverse health consequences and changes in temperature. Temperature records in Malaysia shows potential impacts of climate change on agriculture, forests, water resources, coastal resources, and health and energy sectors. Temperature changes of +0.3°C to +4.5°C, rainfall changes of −30% to +30% and sea level rise of 20–90 cm in 100 years. Malaysia, GHG emissions totaled 144 million tons for carbon dioxide ($CO_2$) equivalent in 1994. Intergovernmental Panel on Climate Change (IPCC) methodologies need to be refined, particularly taking into account local conditions. Table 2.2 shows Industrial sectors that contribution to GHG (MOSTE, 2000).

**Table 2.2:** Green house gas emission impact

| GHG | Amount | Industrial contribution |
|---|---|---|
| CO2 | 67.5%, | Combustion energy sector accounted for 86.7% of total CO2 emissions, landfills (46.8%) and fugitive emissions from oil and gas (26.6%) |
| CH4 | 32.40% | landfills (46.8%) and fugitive emissions from oil and gas (26.6%) accounted for 73.4% of total CH4 emissions |
| N2O | 0.10% | Traditional biomass fuels accounted for 86.4% of total N2O emissions |

In maritime industry, pollution issues focus more on release to water, and safety to avoid collision and grounding. But lately, pollution release to air, soil and biodiversification concerned especially in coastal region are aggressively being questioned. Analysis made by the United Nation Environmental Protection (UNEP) regarding region under coastal treat concluded that Asia have a lot of river runs off to the sea than any other continent. The use of inland waterways and accident aversion could curb green house gases.

## 2.5 Environmental Risk of IWT

On a global scale, pollution is a growing threat to both human health and the environment. Commercial freight transportation, with its almost total dependence on petroleum based fuels, contributes significantly to pollution. Therefore, each form of transportation, as a major energy user, needs to be evaluated both as to the scarceness and future availability of the energy resources that it uses and to its impact on the environment. With each transport mode having its own specific energy use and environmental characteristics, decisions on transport issues, whether short or long term, have inevitable impacts on the environment, which should be clearly weighed before a final decision is made. Recently, three studies were conducted on what happens if cargo movements are shifted from one mode to another, what would be the increases in (i) fuel usage, and (ii) issues related exhaust emissions, and (iii) probable accidents, and traffic congestion. The benefits resulting from these other purposes are as important as the waterway itself. One specific benefit of waterways is that they can interact with nature as a good environmental neighbor. For sustainable IWT, Risk should be analyzed regarding vessel movement, sediment transport for reliable plan of the waterways. Increased bed load transport and consequent downstream output of bed material caused by channel construction for the improvement of navigation.

### 2.5.1 Marine Pollution Risk

On pollution, Sylvia (1995), renowned oceanographers stated that: "It does not matter where on Earth you live; everyone is utterly dependent on the existence of that lovely and living saltwater soup. Wide variety of human activities can affect the coastal and marine environment. Marine pollution could be as a result of operational activities like oil spill, ballast water, garbage, and bunkering fuel. It could be as a result of emission oxide of sulphur (SOx), nitrogen (NOx), Particulate Matter (PM) chloflorochloride, volatile compound (VOC), noise and vibration. These have consequential socio economic impacts to marine ecology. Accidental consequence could be damage, ecological impact and environmental degradation. Global study agent on marine pollution (GSAMP) estimated that about 300—400 thousands of oil entered the world ocean through collision with marine mammal, which then cause propeller injuries, grounding, stranding, loss of oil, hazardous cargo, noxious liquid, noise (GESAMP, 1997). Discharge of pollutants could also be Intentional and unintentional discharge.

## 2.6  Environmental Benefits of IWT

The advantage of using IWT over other mode of transportation has been described by various comparative studies. Advantage range from issues of concerned to human and the environment. Various transportation comparative advantages declare water transport as the safest and most regulated form of transportation. It has fewer accidental spills or collisions than any other mode of transportation (IMO, 1978). Thus, there is little public awareness of the water transport industry outside the river communities that it serves. This can be attributed primarily to the non-intrusive nature of the industry's operations and its impressive safety record associated their location in relation to population centers, as well as levees and floodwalls (USACE, 1980).

### 2.6.1  Energy Efficiency of IWT

The use of energy by different modes of freight transportation has become of increasing concern in setting transportation policy. Energy efficiency is the measure of performance of our system. Be it structure or mobile energy, efficiency is usually measured in one of two ways by comparing how many miles each mode of transportation can carry a ton of freight per gallon of fuel, or by how energy are expended per ton mile. Table 2.3 reveal that IWT provide by far the most energy efficient transportation compare to other modes of transportation.

**Table 2.3:** Energy efficiency (Eastman, 1980)

| Sector | Energy Consumption |
|---|---|
| Road | 28.3 |
| household | 41 |
| Industry | 29.4 |
| Rail | 0.8 |
| Inland navigation | 0.5 |

The future of transportation is closely linked to the future of world of energy. Emission itself is solidly linked to power generation and consumption. Another study shows that the distance that one gallon of fuel can move one tonne is 59 miles by truck, 202 miles by train, and 514 miles by water. Energy efficiency of barge transportation provides environmental benefits (low air and noise pollution) besides the obvious fuel savings (Eastman, 1980).

## 2.6.2 Congestion Mitigation of IWT

Pressure relating to technological change, needs and population has led to high demand for road transportation vehicle. This has led to inconvenient congestion problems and traffic growth in most cities. This has consequentially increasing GHG release from road transportation. The congestion problem is also showing sign of fringing in infrastructure capacity, where traffic demand exceeds supply and leading to delays and safety problems. Negative impacts associated with congestion include restriction of movement of people and goods, wastes valuable energy resources, increases personal trip times, impairs productivity and social tension. (CBO, 1982).

## 2.6.3 Air, Noise and Vibration Pollution Benefit of IWT

Rise in traffic volumes due to urban population and increase mobility has been identified by recent studies to be main contributors to noise levels rise and contamination of air quality. Transportation mode comparative studies have revealed that road transportation is the major offender more than other mode of transportation.

**Table 2.4:** Annual emissions transportation (Jenseb, 1996)

| Emission source | Tow boat | Other transportation | Other mode |
|-----------------|----------|----------------------|------------|
| NOx             | 3297     | 105932               | 433637     |
| HC              | 939      | 198063               | 295124     |
| CO              | 2101     | 980944               | 3852753    |
| SOx             | 462      | 7887                 | 1234395    |
| Particulate     | 198      | 8940                 | 354672     |

**Table 2.5:** Emission reduction potential (Osterreichische, 2007)

| | | Percentage (%) | | | | |
|---|---|---|---|---|---|---|
| | | NOx | PM | FC | Cox | Sox |
| After treatment | SCR (Selected catalytic reduction) | -81 | -35 | -7.5 | -7.5 | -7.5 |
| | PMF (Particulate matter filter) | None | -85 | 2 | 2 | 2 |
| Drive management systems | ATM (Advising temporal) | -10 | -10 | -10 | -10 | -10 |
| Diesel fuel quality/ substitutes | (BD) Bio – Diesel | -10 | -30 | 15 | 65 | ~-100 |
| | BDB (Biodiesel blend , 20%BD) | 2 | -6 | 3 | -13 | ~-20 |
| | LSF (Low sulphur fuel) | None | -1.7 | None | None | ~-100 |
| New engine technology | NGE(Natural Gas Engine) | -98.5 | -97.5 | 4.5 | -10 | -100 |

A study by the Engineering Committee of the International Association of Great Lakes Ports calculated that vessels produced peak noises lower than either those produced by a truck operating under normal conditions or by a standing diesel locomotive as shown in Table 2.4 (Jenseb, 1996). A GHG study data in Table 2.5 shows current performance of new and retrofitting technology that is being considered against the threat of GHG emission (Osterreichische, 2007).

## 2.6.4 Social Benefit of IWT

Trucks and trains operate much closer to populated areas, and release of large amount of pollution and noise to the residence. By contrast, river barges have little impact on densely populated areas. Barge transits are relatively infrequent because of the large tonnage moved at one time. River operations take place in channels away from the shore. The engines of a towboat are usually below the water line, which muffles the sound. Surface traffic, both road and rail, near residential neighborhoods contributes to visual, physical, and psychological barriers that can lead to the fragmentation of those neighborhoods (Dharam et al., 1992).

## 2.6.5 Economic Benefit of IWT

The political and economic changes of nation are a factor that maneuver and created dynamic emerging economy. Economy and environmental analysis could bring assurance to drive the transport policies that promote modal shift. Table 2.6 show estimated throughput at Klang port by (Kader, 1997). The making of IWT requires economic analyses that identify trade growth, and reduce consequential rapid rise in the amount of traffic. Commercial transport in Malaysia corridor has soared growing more than 100% in the last decade, with by far the largest increase registered in road transit. It is expected that Malaysia will continue in this dynamic. Traffic flows could grow correspondingly, cargo projection made for the major Malaysia port is shown below. Cargo capacity is major economic advantage of inland waterway, it is also shows clearly that IWT has a good advantage over other mode of transportation in regards to capacity of cargo.

**Table 2.6:** Projected throughput for bulk and container

| Year | Bulk cargo (Tones 000') | Container (TEU 000') |
|------|--------------------------|----------------------|
| 2000 | 8819 | 1102 |
| 2005 | 11364 | 1407 |
| 2110 | 13909 | 1712 |
| 2015 | 16454 | 2017 |
| 2020 | 19000 | 2322 |

## 2.6.6 Safety Benefit of IWT

Land based transportation mode including road and rail cars are susceptible to accidents. Often the accident result to loss of cargo, especially rail transportation are more vulnerable because shipments typically involving a large number of massive units traveling at high speed in a single line. According to an intermodal comparison work conducted by National Waterway Foundation (NWF), several theories help to explain relative spill and damage frequency. The expected number of accidents is directly related to the number of modal units required to transport a certain amount of tonnage (NWF, 1983).

Collision in waterways falls under high consequence incidents whose data could be incomplete or imperfect or inconstant. This make it difficult to account for dynamic issues associated with vessel and waterways. This necessitates the conclusion to analyse collision base on historical, if not then, probabilistic or simulation methods. It is important to address risk factor linked to collision in waterways and model accident probability for system and subsystem causal factors. For example reliability of propulsion failure model to prevent, control collision and protect the environmental. Developing sustainable inland water transportation requires transit risk analyses of waterways components and relationship between various factors. Risk and reliability based design entails the systematic integration of risk analysis in the design process. This could target risk prevention and reduction as a design objective that support holistic approach to ship and waterways design is needed to enable appropriate tradeoffs for decision making leading to optimal design solutions (Hollnagel, 2009).

Berthing and mooring design is an important part of inland water channel design. Depending on the types of cargo, environmental concern berthing configuration and fender design can be chosen according to energy impact. For berthing and mooring safety, the most severe combination of the environmental loads has to be identified for each component. At a minimum, the conditions of current, tide, vessel loading and wind direction should be considered. The total environmental loads on a moored vessel comprised of the lateral load at the vessel bow, the lateral load at the vessel stern and the longitudinal load and line pretension loads. Design and selection of new fender systems can follow guideline for designs of fender by PIANC (1997).

## 2.7 Critical Review on Risk and Reliability Modeling

Risk analysis involves prediction of hazards to ships, human beings, the environment damage or release. Results of a risk analysis help determine critical accident scenarios that may jeopardize the safety of ships and the environment damage and degradation). Risk in Probabilistic Risk Analysis (PRA) and Probabilistic Safety Analysis (PSA), is defined as the product of the probability of the unwanted event and of the consequences of the event if it occurs. The probability is defined as the number of events per time unit, for example as the probability of the number of collisions per year. The costs describing the seriousness of the consequences might be for example lost human lives in a year or the cost of cleaning oil spills in a year. The objective of risk analysis is to find out what might happen, how probable it is and what are the consequences. Traditionally, casualty statistics is analyzed to get the average accident rates. Statistics give valuable information but cannot be used for required needed purposes or justify current situation. In order to build responsible IWT, it is important to understand need analysis. This can be achieved through examination of the components of system functionality and capability that include major requirement and classification of coastal water transportation system. Figure 2.1 shows question that risk analysis tends to answer.

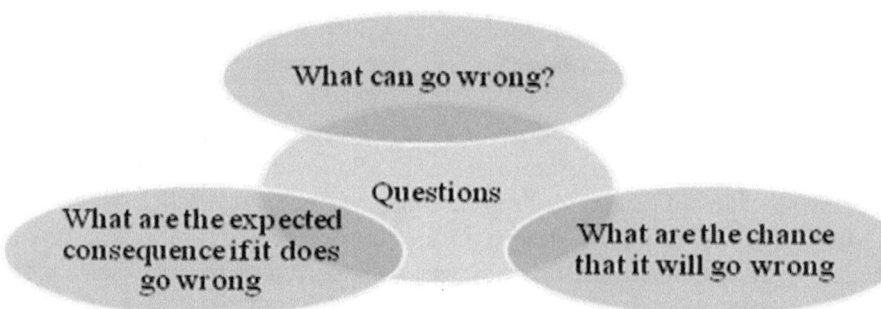

**Figure 2.1:** Risk analysis requirement

Qualitative risk identification work should be followed by scientific and system systematic based quantitative risk analysis that include risk ranking, limit acceptability and generation of best options for development sustainable IWT. Risk and reliability process should consider safety and environmental risk, high goal objective and life cycle evaluation of the system throughout for the development of reliable and cost effective IWT as sustainable alternative green technological choices. In maritime industry risk assessment and analysis has been instrumental to make reliable decision related to prediction of flood, structural reliability, intact stability, collision and grounding, fire safety contemporary time has seen risk assessment optimization using scenario based assessments, which consider the relative risks of different conditions, scenarios and event (DOC, 1974). HAZOP technique is commonly use for qualitative analysis HAZID, it involved use Guideword: i.e. No pitch, No blade, Description: i.e. no rotational energy transformed, object in water break the blade, Causes: i.e. operation control mechanism and Safety measurement to address implementation of propeller protection such grating, jet.

Failure Modes and Effects Analysis (FMEA) is a method to analyze potential reliability problems in the development cycle of the system components. Making it easier to take actions to overcome such issues, and enhancing the reliability through design. FMEA was developed in the late 1940s for military usage by the US Armed Forces (Moderras, 1993). Today it is widely used in most industry. It has been employed in the following areas like the aerospace industry during the Apollo missions in the 1960s and the US Navy in 1974 developed which discussed the proper use of the tool.

Today, the FMEA is universally used by many different industries. There are three main types of FMEA in use today system FMEA: concept stage design system and sub system analysis, design FMEA: product design analysis before release to manufacturers, and process FMEA: manufacturing or assembly process analysis. Risk of pollutant is ranked according to toxicity and associated abnormal development, growth and reproduction of organisms, as well as lethal reaction, due to immediate or prolonged exposure to contaminants. The accumulations of toxins have the potential to contaminate human food supplies. Analysis of risk should be ongoing throughout the channel operation life. Transport contaminants in the environment do not always remain in one place, but become transported throughout the environment through several different processes. In areas of contamination the chemicals will tend toward an equilibrium concentration between solid, liquid and gas phases. Tidal flow regimes, vessel traffic, and dredging activity tend to cause physical resuspension of the sediment through redistribute contaminants (Peddicord, 1997).

Weighing of deductive balancing work requirement for reliable and safe IWT through iterative components of all elements involved inclusion of social, economic, health, ecological, technological consideration in the risk process. Concerned related to incorporating other use water resources into the system through best practice technology, beneficial disposal of sediment, mitigation for environmental impact, continuously management, monitoring, compensation for uncertainty and preparation for future regulation is also important for the risk and reliability work. Probabilistic and stochastic risk analysis and concurrent use of virtual reality simulation that consider the broader impacts of events, conditions and scenarios is useful for the best result. Selective tool should reflect geographical and temporal impacts, risks of conditions under study, sensitivity and contingency as well as use of "what if" at preliminary risk assessment stage to deal with as much possible remnant reliability answer to uncertainty in the system.

## 2.7.1 Risk Mitigation Best Practice

Risk mitigation should target environmental risk prevention and control measures options. It involves elimination or reduction of risk and those that cannot be reduces to be put under control. Quite a few methodologies exist that consider risk as part of assessing safety. Most methods refer to risk has probability of occurrence of an unfavorable event multiplied by the consequences should such event occurs. A round table workshop could also consider to identified strengths and weaknesses of the approached adopted. The outcome of risk work could decide provide decision support on the next immediate action or suggest future research on needs relating to each application. Best practice options are normally transformed into guidelines (Fuji, 1974).

### 2.7.1.1 *Last Work on Risk and Reliability Based Methods*

Risk and reliability based analysis is a mathematical approach to engineers and predict the risks of accidents and give guidance on appropriate means of minimizing them. Existing work on marine risk based design are shown in Table 2.7.

**Table 2.7:** Best practice of risk and reliability based design

| Institution | Studies |
|---|---|
| The Norwegian Petroleum Directorate | Guidelines on how to apply risk analysis to meet its regulations |
| UK Health & Safety Executive | Guidance on risk assessment in the context of Offshore Safety Cases |
| Canada-Newfoundland Offshore Petroleum Board | Guidance on installation Safety Analysis to help operators meet its regulations |
| American Petroleum Institute | Recommended practice for design and hazard analysis offshore production platforms. |
| The UK Offshore operators Associations | Procedure for the conduct of formal safety assessment of offshore installations, with very brief coverage of hazard assessment. |
| Pitblado & Turney (1995) | Introduction to QRA for the process industries, |
| Aven (1992) | Discussion of offshore QRA, focusing in particular on reliability analysis. |
| Crook (1997) | Qualitative review of recent technical and regulatory developments in the field of safety against fire, inherently safer design, and human factor. |
| Brian Veitch | Rescue and evacuation from offshore platform |

Nevertheless, while it uses scientific methods and verifiable data, the method is immature and highly judgmental technique and its results have a large degree of uncertainty. This situation can be avoided by combining the predictive method with deterministic method, an approach used in this model. Despite this, many branches of engineering have found that risk based method can give useful guidance for sustainable system design. Risk base method is not the only input to decision making about system safety and environmental protection. Other techniques based on experience and judgment may be appropriate as well complement this method. Risk assessment does not have to be quantitative, and adequate guidance on minor hazards can often be obtained using a qualitative approach (Vose, 1996).

## 2.7.1.2 *Computer Based Risk and Reliability Analysis*

Early risk based method has been handled with manual calculations based on written documentation, typically supported by hand held calculators. But the approach is suitable only for very simple quantitative risk analysis or for checks of more sophisticated work. Its strengths are flexibility and economy of effort in simple work. Its weaknesses are difficulty in handling large numbers of events and updating after changing inputs, and the variable quality documentation from different analysis. Computer spreadsheets have been used extensively in recent risk studies. They are also widely used as a computing environment for simple consequence models. Spreadsheets are controlled by macro commands, allowing them to function like complete computer programs. The strengths of spreadsheets are their low cost, flexibility of calculation and presentation, minimal training requirements, and easy portability from one study to the next (Murphy, 1996).

Computer programs are mainly used in risk work as single-issue stand-alone models for consequence calculation, fault-tree analysis, and theoretical frequency models for specific events. In this form, they can be combined with manual calculations, spreadsheets or more comprehensive software to produce overall risk results. Spreadsheet and computer modeled are used for development, analysis and validation of this model. Examples of risk based software are Comprehensive offshore quantitative risk analysis software has been developed for Frequency consequence (FC) analysis, The Offshore Hazard and Risk Analysis (OHRA) Toolkit is a graphical tool for structuring an offshore risk analysis, PLATO is a software system for offshore risk analysis which performs the entire risk calculation from definition of system, RELEX and ITEM are other risk software that are popular for risk and reliability based design and MINITAB is a tool use for validation (Roach, 1998).

## 2.7.2 Benefit and Limitation of Risk and Reliability Based Method

Rampant system failure and problem related to reliability has brought the need to adopt new philosophy to design, operate and maintain system base on top down risk and life cycle risk model. Election of alternative way to mitigate challenges safety and environmental risk deserve holistic, reliability and risk engineering system based approach. iterated in his work that the approach provides the benefits that Lead to alternative improvised design and concept development for evaluation of risk reduction measure, that involve transparency of decision making process and use of systematic tool to study complex and dynamic system and their interaction between discipline, risk, impact and valuation of system, that facilitate proactive approach for system safety that improves current design practice, management that facilitates total system approach that touches on risk contributing factor in system work and others establishment of systematic rule making, limit acceptability and policy making for development and analysis of transportation system.

Major inherit draw back associated with risk based model are Lacks of historical data (frequency vs. probability, expert judgment), Linking system functionality with standards requirement during analysis (Total safety level and individual risk level, calculation of current safety level) and Risk indices and evaluation criteria (Individual risk acceptance criteria and sustainability balance) (Cornel, 1996).

## 2.7.3 IWT Capability and Standards Requirement

The process sustainable risk and reliability analysis should be carried out with hybrid the use of goal based and risk based design towards beyond compliance and system matching. This is incorporated in the risk process in order to capture the gap between standards and system functionality. It is clear that the shipping industry is over killed with rules and recent environmental issues have potential to initiate new rules and make firms to selectively adopt beyond compliance policy that are more stringent than the required law. Beyond compliance policy is mostly intra firm process which could be power based or leadership. It draw insight from institutional theory, cooperate social performance perspective, and stakeholder theory that relate to internal dynamic process.

While external forces create expectation and incentive for manager, intra firm politics influence how managers perceive, interpret external pressure and act on them. Policy towards beyond compliance fall the under whether they are now required by law but they are consistent with profit maximization or requirement by law and firm are expected to complied. Towards system sustainability and reliability, it is preferable to use stochastic and probabilistic methods whose prediction could help improve the existing methodology. This involves formal system based approach that will cover more uncertainty and compliment flaws in other historical data dependant methods. Beyond compliance towards meeting required safety level, life cycle and environmental protection required systematic employment of hybrid analysis of system functionality and standards requirement for reliable and efficient IWT.

Sustainability remains a substantial part of assessing risk and life cycle of IWT. However, they are very complex and require long time data for accuracy. Environmental impact assessment procedure is laid out by various environmental departments. And it will continue to remain similar except that the components of risk area cover are not fully scientific method employment make a difference in this. Also uncertainty to sustain a particular system may also be different. Environmental impact assessment has been a conventional process to identify, predict, assess, estimate and communicate the future state of the environment. With and without the development in order to advise the decision makers the potential environmental effects and risk of the proposed course of action before a decision is made is necessity by using scientific, formal system stochastic process (Peterson, 1997).

### 2.7.3.1 Risk Based Design Functionality

FSA is improvised version of Environmental Impact Assessment (EIA) for international transportation maritime system. It consideration whole system, community participation, expert rating, cost benefit analysis and regulatory. It is concerns with core part of the philosophy leading to reliable decision making, sustainable system design, operation and maintenance. GBS deal with life cycle issue of the system. It is established to compliment FSA towards sustainable system design model. In risk analysis, serenity and probability of adverse accident or hazard are deal with through systematic process. That quantitatively measure, perceive risk and value of the system using input from all concerned waterway users and experts (Funtowicz et al., 1992).

RISK = Hazard x Exposure                                    Eq. 2.1

Risk is an estimate on probability that will lead to certain consequence in term of damage, oil outflow, toxicity etc. Hazard constitutes anything that can cause harm in term of chemicals, electricity, natural disasters. Severity may be measured by number of people affected, monetary loss, equipment downtime, area affected, and nature of credible accident. Risk management is the evaluation of alternative risk reduction measures and the implementation of those that appear cost effective where zero discharge is expected to be equal to zero risk.

### 2.7.3.2 Risk Based Design Standard in Maritime Industry

According Sirkar et al., (1997), past rules by IMO are prescriptive conservative and rigid. Risk based design based criteria allow better flexibility for innovative designs at optimum maintaining safety and pollution performance. IMO guidelines are for approving alternative designs which provide a probabilistic method to assess consequence of accident under FSA and GBS. FSA was introduced by the IMO as "a rational and systematic process for accessing the risk related to maritime safety and the protection of the marine environment and for evaluating the costs and benefits of IMO's options for reducing these risks" (see FSA Guidelines in MSC/Circ.1023, MEPC/Circ.392, 2002). Previous experience in the use of FSA includes Psaraftis, (2002); In 1997, when the IMO reversed its prior position to require Helicopter Landing Areas (HLAs) on all passenger ships even before the relevant regulation had come into effect. May 2004 decision of IMO not to impose mandatory double hulls on bulk carriers was based on an FSA study, even though the IMO's prior opposite view was essentially based on other studies that used the same method ii) "International Collaborative (IC) FSA Study", managed by the United Kingdom, recommended the mandatory construction of Double Side Skin (DSS) for bulk carriers (MSC 76/5/5).

### 2.7.3.3 Accident Scenario Assessment

Based on balanced consideration of the criteria for system design output from the HAZID and with reference to the statistics of accident figures for barges, generic accident scenarios are common to all ships in waterways are (Kite, 1996): Collision scenarios represent on average 35% of all relevant initial causes in the casualty statistics, Contact is most collision scenarios that occur in fairways and sea, Grounding scenarios represent on average 20% of all relevant initial causes, Fire and explosion scenarios represent on average 10% of all relevant initial causes.

Baseline data include waterways river profile, associated technocrat and environment. Further data regarding navigation requirement including river profile, channel and vessel requirement which represent functional requirement for high level objective for risk and goal based design which are assessed. In risk and reliability based design, standards requirement that complement system functionality analysis for safety and mitigation of environmental hazard of vessel and waterway is assessed. This provides groundwork for preliminary hazard analysis (PrHA), HAZID, FMEA for collision scenario, as well as analysis of the whole system and subsystem, Eftratios (2005).

### 2.7.3.4 Collision Risk Scenario Modeling

Collision is the structural impact between two ships or one ship and a floating or still objects that result to damage. Collision is considered infrequent accident occurrence. Its consequence can be economically, environmentally and socially significant (Fuji, 1974). Observation on Langat River accident data and concern about yearly increase in collision warrant urgent need for establish appropriate sustainability balance between the interests of safety, environmental protection and the economics. Collision and grounding accidents remain main accident in waterways despite efforts to prevent them. Recent years has seen increasing demand for safety at sea and protection of the environment and it has become very imperative to predict an accident, assess its consequences and ultimately minimize the damage of an accident to ships and the environment. The main spectrum of issues related to collisions and groundings has been critically reviewed in this. Recent technological development and research work has enhances the understanding of this complex problem. It is better to use mix of deterministic, probabilistic, or semi-probabilistic for criteria establishment methods. Statistic of accident has been one of the main areas widely research and being documented by MAIB, IMO, DnV, Brown et al., (1996) did extensive research regarding framework for collision of ship and development of evaluation approach for consequence of collision and compilation of structural performance of different designs.

Navigation accident comes in form of passing vessel collisions (Agiwara, 1983), incoming vessel collisions (Lee, 1999), and structural failures on mobile and fixed platform. Accident scenarios represent the events to be considered probability for accident scenario in his research hereto collision. This can be analyzed through the following methods; statistics from historical data, associated probability of occurrence, Expert opinions, and Risk analysis Brown etal. (1996). A typical collision event modeling process requires:

i. Accident events fail study, followed by investigation
ii. Extensive system description including input, output and human machine variables
iii. Description of events that leads to the accident, responsible contributing party the event including, management
iv. Outcome of analysis is followed by suitable risk control options
v. The environmental impact
vi. Financial consequences to local communities close to the accident,
vii. Damage to coastal or offshore infrastructure, for example collision with bridges
viii. Unplanned passage is possible, but effectively manageable, frequently preceded by related events.

Table 2.8 describes the previous risk method that has been employed in maritime industry. Combination of stochastic, statistical and reliability method based on combination of predictive, goal based designed, formal safety assessment methods and fuzzy, method using historical data's of waterways, vessel environmental and traffic data as necessary could give a better result in risk work (Kite et al, 1996). Risk based design can be modeled by conducting the following analysis Raiffa (1969):

i. Risk identification, involve fact gathering, assessment and establishment of a realistic representation of the navigational activity that answer question relating to how and why accidents occur.
ii. Risk analyses; require scientific based quantitative analysis on selected collision scenario.
iii. Damage estimation involve the consequences of structural.
iv. Priotisation of risk level.
v. Risk iteration for sustainability analyses, to define the risk, consequence and cost of mitigation.
vi. Mitigation through acceptance criteria establishment, evaluation of preventative and protective measures relating to how mitigation can be achieved.
vii. Repriotization of exposure category mitigate risk or consequence of events.
viii. Reassess high risk events for monitoring and control plan.
ix. Recommendation, implementation and monitoring.

Table 2.8 shows previous waterway accident risk model.

**Table 2.8:** Previous waterways accident risk model application

| Model | Application | Drawback |
|---|---|---|
| Brown et al (1996) | Environmental performance of tankers | Damage analysis deal only with oil spill |
| (Sirkar et al (1997) | Consequences of collisions and groundings | Difficulties on quantifying consequence metrics |
| Brown and Amrozowicz | Hybrid use of risk assessment, probabilistic simulation and a spill consequence assessment model | Oil spill assessment limited to use of fault tree |
| Sirkar et al (1997) | Monte Carlo technique to estimate damage + spill cost analysis for environmental damage | Lack of cost data |

| IMO (IMO 13F 1995) | Pollution prevention index from probability distributions damage and oil spill. | Lack (Sirkar et al (1997)). rational |
|---|---|---|
| Research Council Committee(1999) | Alternative rational approach to measuring impact of oil spills | Lack employment of stochastic probabilistic methods |
| Prince William Sound, Alaska, PWS (1996) | The most complete risk assessment | Lack of logical risk assessment framework (NRC (1998)) |
| Volpe National Transportation Center (1997) | Accident probabilities using statistics and expert opinion. | Lack employment of stochastic methods |
| Puget Sound Area USCG (1999) | Simulation or on expert opinion for cost benefit analysis | Clean up cost and environmental damage omission |

Risk is defined as the combination of probability of occurrence and severity of consequence the start with HAZID, mostly in exclusively relies on a group of experts. In order to fully understand the advantages and disadvantages of expert opinion, a closer look at it with both mathematical and behavioral approaches is necessary. Experts should identify hazards using any method and provide their rankings for each hazard. The expert's idea can contribute to the development of an international approach with a view to ensure that, in the future. The potential source of all problems is the fact that most studies avoid probabilistic modeling and use casualty historical data and frequencies. Collision work that consider collision frequency are Pedersen (1993), Fowler (2000) and Macduff, T. (1974). Pedersen (1997) consider the ship in the waterway 1 approaches the ship in the waterway 2 with relative velocity that is formulated in the following way:

$$V_{ij} = (V_i)^2(V_j)^2 - 2V_i V_j cos\,\theta \qquad\qquad \text{E.q 2.2}$$

Where: Vi is the velocity of vessel in ship class i in the waterway 1, Vj is the velocity of ship class j in the waterway 2 and θ is the angle between incoming directions of vessels, Li is the length of vessel in ship class i in the waterway 1, Lj is the length of vessel in ship class j in the waterway 2. Equally, B is the width of vessel. Ships are classified to groups by their type and length. Figure 2.2 shows ship and waterway interaction.

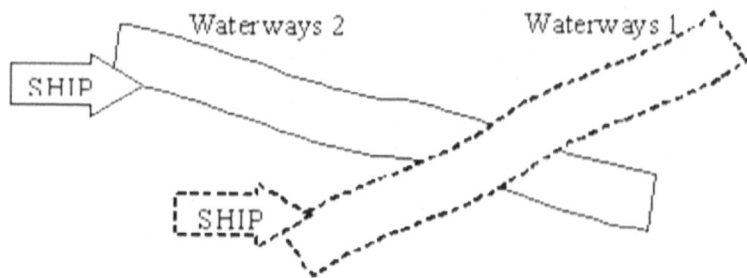

**Figure 2.2:** Ship and Waterway interaction

The work considers gives the geometrical collision probability meaning ships that would collide if no aversive maneuvers were made. Drifting ships are not taken into consideration in their mode. Fowler (2000) collision work powered collision signifies that collisions frequency of encounters is calculated assuming that traffic movements are uncorrelated. The model evaluates the frequency of ship-ship powered collisions. The collision candidates of their model are not all real collision candidates as all of the ships in encounter situation of the model would not collide even though the crews of the involved ships were asleep and did nothing. Each encounter is multiplied by the probability of a collision per encounter, to give the collision frequency, *fco*.

$$fco = nco\ (pc\ P\ co,\ c + pf\ Pco,\ f), \qquad\qquad\qquad \text{E.q 2.3}$$

Where *pc* and *pf* are the probabilities of clear and reduced visibility. *Pc*, and *P*, are the probabilities of collision given an encounter in clear or reduced visibility. The critical situation denotes that two ships are crossing within half a nautical mile of each other. Encounter frequency at a location estimated by a pair-wise summation across all shipping lanes. Calculation may either be done to all vessels or specific ship types.

Macduff (1994) collision work involve situation whereby a ship is expected to approach a shipping lane on a course that makes an angle θ with the lane. Where the mean free path of a ship is defined as the distance which the ship can proceed, on average, before colliding with one of the vessels navigating along the main shipping lane. It is based on molecular collision theory where all vessels are supposed to navigate with the average speed of ships. A fault tree is a logical diagram that determines the probability of hazard using Boolean logic to combine a series of lower level failures such as navigator asleep and radar failure. The geometrical collision probability is represented by:

$$Pc = XLD\ sin\ \theta \qquad\qquad\qquad\qquad \text{E.q 2.4}$$

Where D is average distance between ships (density measure, in miles), X is the actual length of path to be considered for the single ship (nautical miles) and L is average vessel length, Fowlers (2000) work did not describe how to calculate the described geometrical collision probability. The Fowler (2000) calculated collision probabilities for different ship groups, but did not consider the fact that ship dimensions should be integrated in causation factor as the number of encounters does not change according to the ship size. Macduff (1994), traffic use continuous uniform flow and not as single ships with their individual characteristics as type, dimensions and velocity. Real marine traffic of ships of equal size moving at equal velocity at equal distances from each other which makes his work imprecise, also his work did not take into consideration differences in between ships. Pedersen defines collision candidates geometrically that take the characteristics into consideration in the model. He calculated collision probabilities of different ship groups where collision candidates would really collide with each other if nothing was done to prevent the accident. The model of Pedersen(1993) seems to be the most suitable one for estimating collision risk frequency.

Consequences involve ship loss to human losses or environmental harm. These methods differ in methodology, in theory, in complexity of calculation, and in the cases to which they apply. The total accident problem can be take two approach, firstly external dynamic that deals with the energy released for dissipation in damaged structures and the impact impulse of a collision or grounding which is used in this research and the second approach is the internal mechanics, which deals with the strength or resistance of ship structures in an accident consequences is modeled using the Potential Loss of Life (PLL) is number of fatality / shipyears (Devanney, 1967). Figure 2.3 shows length of collapse obtained by Klanac et al (2005).

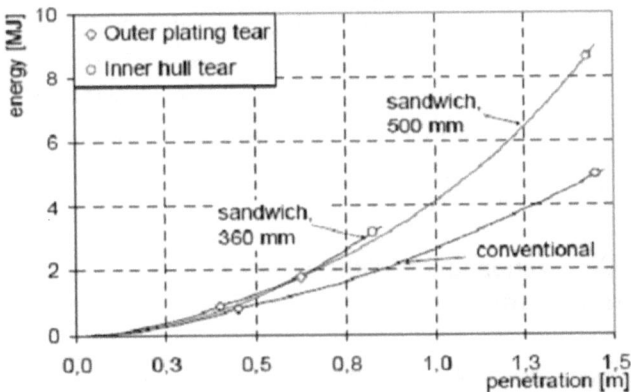

**Figure 2.3:** Energy absorption and length of penetration

The formulae that are derived from theoretical analysis correlate empirically the energy absorption with the volume of the damaged ship structures. The formulae are more rational in mechanics to incorporate in a systematic evaluation system can also be derived through extensive theoretical analyses. The analytical methods generally capture the characteristics of a damage process, and employ theoretical formulae for structural components if required. They provide insights into both global and local levels which are more advanced tools which designers can use for analyses (Wang et al., 2001).

The method can be applied to a wide range of accident situations which include; head-on collision on rigid walls, ship–ship collision, ship–platform collision, ship–bridge collision, bottom raking, and stranding. The analytical methods can also be used for structural crashworthiness which is a proven practice in the automobile industry. This involves identify primary damage patterns of structural components according to observation of actual damage, develop idealized theoretical models and derive theoretical formulae to capture the main features of the damage patterns, establish global models for the entire damage process of the hull; and combine the global damage models with formulae for individual structural components.

A ship may collapse after an accident because of inadequate longitudinal strength. The hull girder section modulus is a well-accepted parameter measuring the longitudinal bending strength. Practically, hull girder section modulus can be used to quantify the hull girder strength in case of an accident Wang et al., (2001). To minimize damage and causality it is important consider adequate stability, containment of cargo from spilling, resistance to the accidental loads, and sufficient residual strength. (IMO, 1994) has established a probabilistic methodology for evaluating the oil outflows from a damaged ship. The intensive calculations use rational approaches while keeping the same assumptions for damage extents and similar methodology for oil trapped in double bottom (Sirkar et al., 1997). As it has been mentioned before, FSA was conceived as a tool to provide a transparent decision-making process, clearly justify proposed measures, and allow comparison of different options. The total terminal damage fraction is then:

$$D_F = \sum D_m * C_m / C_P \qquad\qquad \text{Eq 2.5}$$

Where: $D_M$ fraction of module damaged, $CM/C_p$ cost of module as fraction of terminal cost. The annual damage fraction is then:

$$ADF = \sum_{scenarios} F_S * D_F \qquad\qquad \text{Eq 2.6}$$

Oil Spill is another alternative way to account for damage estimation by:

$$ASR = \sum_{allscenarios} F_S * S \qquad \text{Eq 2.7}$$

Where, the quantity of oil spilled, S, is estimated for each scenario.

The frequency and consequence assessments provide the risk associated with the different generic accident scenarios. This can be summarized to estimate the individual and societal risks pertaining to IWT operations and design. (Otsus, 2007).

### 2.7.3.5 Data Requirement

The accident data are quite scares, that the probability of such an accident would be zero. In addition, there are some problems with statistics, like data about accidents that is stored differently in variable databases. As a result, all information is difficult to put together. These are the essential reasons why rational risk calculation is needed.

### 2.7.3.6 Historical data

Kite-Powell et al., (1996) work revealed that unbiased statistics are the most reliable sources for identifying typical and critical accident cases. Thus available statistics suffer are proned to inconsistencies and incompleteness. Mostly, conditions surrounding an accident like ship speed, ship's loading condition, environment condition remain most that are not mostly recorded. In order to minimize misrepresentation of data, it is important o bear in mind that statististics are activities that may not represent present situation; therefore Hybrid of combining limited historical data with risk analysis techniques and experts' opinions provides a better basis to determine realistic and critical accident scenarios.

### 2.7.3.7 Experts' Opinions on Collision

Identified limitations in data collection poised hybrid use of historical and stochastic analysis to support the available historical data. Future data collection effort can open opportunity to cover uncertainty gap, improve analysis and understanding of accident risk. Major data problems include environmental data: limitations are associated with potential change in real time, Port specific data: information about safe transits counts categorized by flag, vessel type, vessel size, with tug escort and piloting information, taken at hourly by port authority, surveys and chart data, it is important to compare conventional cartographic uncertainty and with new technology. Additional uncertainty include vessel casualty data, inherent problem with causality data are missing entries, duplicate entries, and inaccuracies, location of accident, Lack of recording of location of accidents in theory expected to be to tenths of minute's latitude and longitude for accuracy, stability information, No data actual draft or trim of vessels at the time the accident happen, no actual water depth at the time, Lack of accident related to the environmental data and Lack of quantification in the use of tugs and present of pilot.

### 2.7.3.8 Baseline Data Requirement

Waterway economic and technical data: IWT development works have safety and environmental implications. There is the need to carry out safety and environmental assessments and analysis before decisions are made on complex and dynamic system like IWT. Figure 2.4 shows IWT channel and vessel dimension.

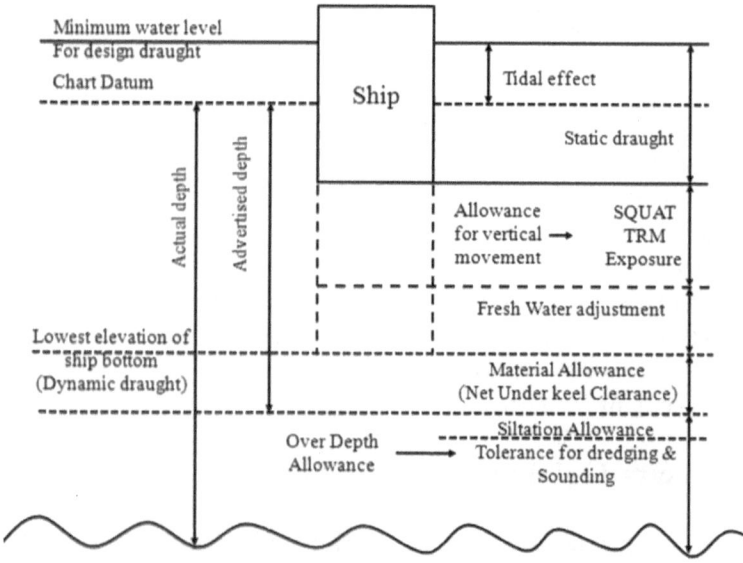

**Figure 2.4:** Channel and vessel dimension

The process must include culture of integrated planning of navigation and environmental safety optimization needed to minimize legal costs, delays and sometimes unstable outcomes. Figure 2.3 show relevant baseline data for collision in waterways risk. Conventional environmental data that need to be taken into consideration in the risk process for collision model are derivative for improvement of maritime accident data collection, preservation and limit acceptability. The factors relating to the following information are also important in collision risk process record ports of incidents, record wind speed and direction, visibility, water level, current speed and direction, etc, correct erroneous and duplicate entries (e.g. location information), record data on actual draft and trim, presence and use of tugs, presence of pilot, record types of cargo and vessel movements, report barge train movements as well as individual barges, and improve temporal resolution like transits by day or hour.

The main IWT risk contributing factors data can fall under (Roeleven, 1995): Vessel characteristics, maneuverability capability of vessel could determine probability of accident. Maneuverability data is difficult to acquire in waterways that has no Automatic Identification System (AIS) and Vessel Traffic System (VTS) in place. Therefore analysis relies on derivative from vessel type and size. Barge trains are, in general, likely to be less maneuverable than ships. Topographic difficulty of the transit, the number of bent in the channel also adds to channel complexity which needs further consideration including environmental condition. Water level, accident due to tide are much linked to grounding. low water level, or large (negative) errors in the tide forecasts used by vessel operators, could be a risk factor for groundings., this include quality of operator's information, including information about currents, tide levels, and winds, uncertainty in surveys and charts., real time environmental information and operator skill.

### 2.7.3.9 Technical Data Requirement

Quantification of channel requires quantifying the depth that paves way for dredging requirement. Iterative process for allowance and discounting of impacts to channel during operations and during construction should also be incorporated. Waterway channel involve the sizing of vessels that will transit a waterway. Conducting maintenance dredging capacity, sediments output estimates with clear objective to reduce channel delay and accepts big ships, need to be done in environmental sustainable manner. Characteristics attached to channel are layout, channel cross section channel width and deepening including dimensioning component. Those factors feed into the determination of the dimensions and specifications of channel characteristics. This includes vessel traffic, characteristics environmental factors location and characteristics of features such as bridges and economics along with many others.

Navigation, coastal and geotechnical engineers have a very pronounced problem in past design due to reactive behavior to accident and calamity. This is also due to situation where engineers deal with high level of uncertainty by conservatively assigning or specifying much larger capacities than the projected demand or vice versa. This ratio of capacity to predicted demand is the classical safety factor approach, which requires significant experience levels to factor right. For example care should be taken for calculating allowance when ships enter channels and ports in brackish or fresh water. Ship draft increases due to the lower water density. The draft increase between ocean and fresh water is approximately 2.6 percent (DON, 1984). Figure 2.5 shows IWT effect of wave on ship motion.

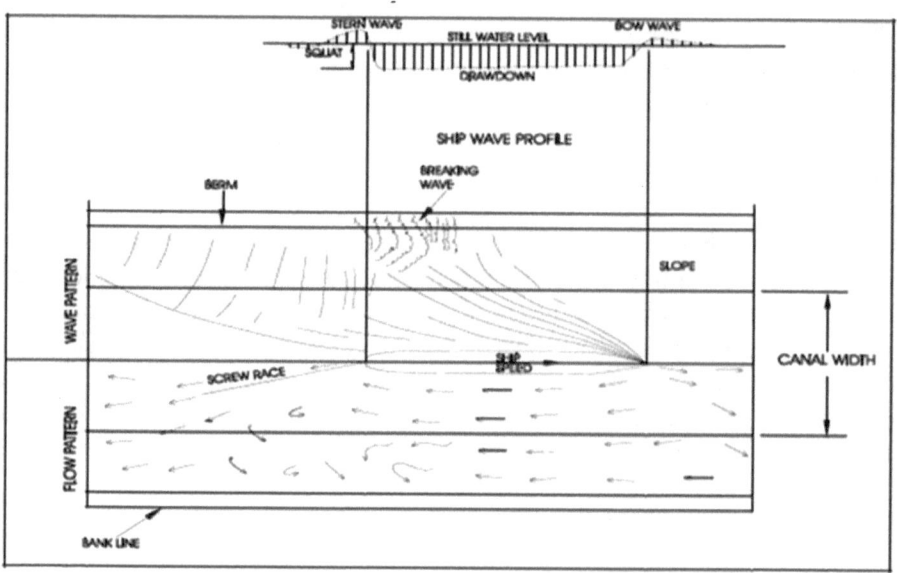

**Figure 2.5:** Effect of waves on ship motion

In practice, a maximum depth allowance of 0.3 m (1 ft) may be included for this effect. PIANC recommend 1 to 1.4 for shallow water and under keel clearance of 40%. Channel depth for both deep and shallow draft navigation may be determined by figuring a depth increment for each of the important factors affecting vessel under keel clearance requirements and adding those to the design vessel draft. Waves are important to ship response in harbor fairways or entrance channels. Ocean waves are classified according the period and origin of the waves as shown in Figure 2.5 (PIANC, 1997). Vessel vertical motion in response to waves must be considered in design of channel depth at exposed locations. Wave effects tend to increase as wave height increases and decrease with longer vessel length. Maximum vessel response occurs with wavelengths approximately equal to vessel length. Most deep draft ships are relatively unaffected by very short period waves but respond become high when periods are longer than around six to eight second. Physical model studies to aid in probabilistic navigation channel design are also used for evaluation (PIANC, 1992).

Direct field measurements of ship motion are valuable for channel design studies, but extreme conditions controlling design are not easily captured. The following are example parametric studies that could be done for necessary wave compensation include analytic studies, using strip theory or other theoretical calculation methods as developed by naval architects, interactive, real time ship simulator studies, physical model studies, using radio controlled free running scaled ship models with wave response measurements, and direct, onboard ship measurements while transiting through the entrance channel (Wallace, 2000). River classification system is necessary to ensure the orderly, efficient control and maintenance of waterways. An inventory of existing infrastructure and transportation information must be prepared as the base of a sound classification system. Classification adopted by European Conference of Ministers of Transport (ECMT) is shown in the Table 2.9. Classification of European inland a waterway provides an example of principles for channel design characteristics. River Langat can fall under class V. Channels are designed to accommodate both the type of vessels and the level of vessel traffic that are forecasted to use a given channel. There are no guarantees that the forecast will accurately predict actual usage. The risk of avoiding vessel size mismatch with channel size should be controlled at all levels for safety, reliability and efficient IWT.

**Table 2.9:** IWT classification (Boil, 1967)

| Class | Type | Carrying capacity | ECMT classification | | | |
|-------|------|-------------------|------|--------|-----------|-------------|
| | | | Beam | Lenght | Air draft | Water draft |
| I | Small barge | 300 | 5 | 38.5 | 3.55 | 2.2 |
| II | Campeenar barge | 600 | 6.6 | 50 | 4.2 | 2.5 |
| III | Doctmu –Ems | 1,200 | 8.2 | 67 | 3.95 | 2.5 |
| IV | Rhi – Hern | 1,350 | 9.5 | 80 | 4.4 | 2.5 |
| V | Large Rhine | 2000 | 11.5 | 95 | 6.7 | 2.7 |

Vessel requirement: Types of Britain's boats and barges for IWT are as follow (Afifi et al., 1999) i) short boats: 40 to 50 tones boat use to carry coal, oil, gravel etc., ii) Narrow boat: 20-30 tones boat use for narrow canals, compartment boats, available on Aire and Calder navigation for carrying coal, iii) Fly boat: express boat on most canals to carry cargo and passenger. Application of new technology in IWT for several types of barge ranging from self-propelled to Lighter Aboard Ship (LASH) is currently gaining ground. As ships are getting bigger, there has been technological change link for safe maneuvering and controllability. Maximum navigational draft is the extreme projection of the vessel below waterline when fully loaded which is needed for navigation channel depth. Mean draft is preferred for hydrostatic calculations. Block coefficient is the vessel shape represented in terms of simple parameters known as form coefficients which represent the fully loaded ship. Values of the block coefficient normally range from around 0.4 for tapered-form, high speed ships, container ships and passenger ferries, to 0.9 for box shaped, slow speed ships, such as barges, tankers and bulk carriers. Limiting speed in channel remains a critical part of operational maintenance work. Mitigation design features to reduce the risk of marine causalities and improving slow speed maneuverability including twin screws and rudders (IMO, 2003). Some ships are also equipped with bow thrusters which aid in control, especially at low speeds. Often, one or more tugs are needed to assist ships in some phases of entering and leaving a port. Vessel operations during navigation channel work are required to enhance safety, efficiency, and productivity of waterborne commerce in ports and harbors which embody public recreational access. Important navigation system related to vessel, port and harbour operability in channel maintenance work is water engineering, traffic engineering and vessel hydrodynamic (Brebbia, 2000).

## 2.7.4 ALARP, Risk Control Option and Acceptability Criteria (RAC)

RAC gives overall risk reduction limit identification and preliminary recommendation. In order to assess the risk as estimated by the risk analysis, appropriate risk acceptability criteria for crew and society should be established prior to and independent of the actual risk analysis. The overall risk associated with inland water should concentrate on the reduction desired areas within As Low as Reasonably Possible (ALARP) influence diagram which leads to cost effective risk reduction measures that could be sought for required system solution (Lind 1996). Figure 2.6 shows ALARP requirement (Vose, 1996).

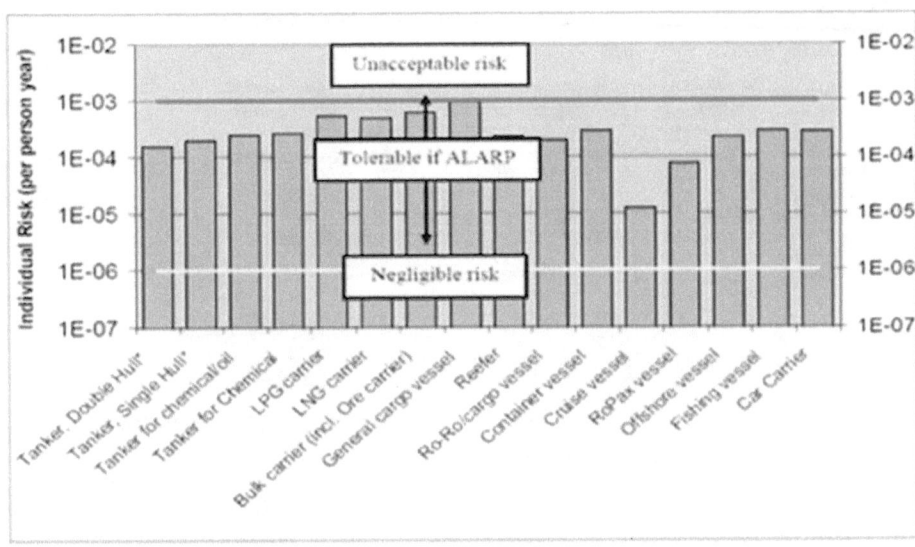

**Figure 2.6:** ALARP risk acceptability

Risk acceptability or performance standards are a key aspect of modem safety management. They are criteria for measuring whether the safety measures in waterways are performing or capable of performing their intended role in minimizing risks. While risk acceptability criteria perform a similar role in helping to audit the overall risks on the installation, performance standards address individual systems, items of equipment or safety procedures in place. Risk acceptance criteria are often given in the form of bounds on the annual probability of failure in dependence of the consequence of failure measured as monetary costs or lives lost. Examples of such bounds together with indications of domains of experienced fatalities or lost capital are referred in (Bea 1990).

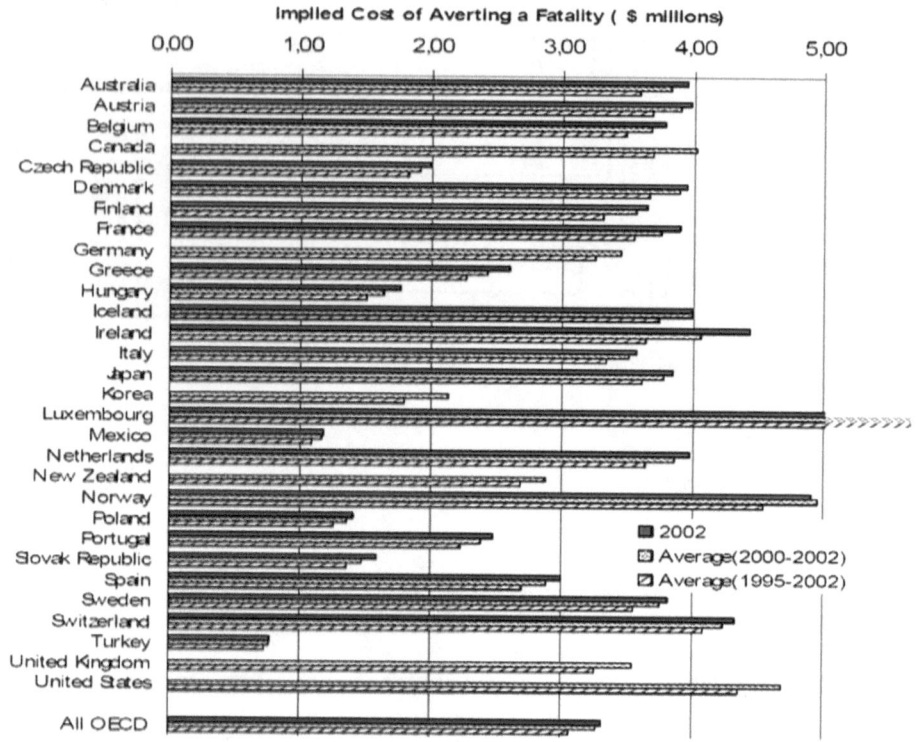

**Figure 2.7:** Implied cost of averting a fatality

The value of human rest, pleasure, felicity, etc. is outside rational reasoning and can for use in decision analysis involving anonymous people be set directly or indirectly only through a decision made by the political authorities. From the decision maker's point of view (that is, from the point of view of the owner of the capital producing operation) the concern is only the cost of the compensation (and the loss of reputation and perhaps goodwill) that matters. To prevent cynical exploitation of human lives for large benefits, the public is forced to impose restrictions that may require suitably enhanced utility losses to be used in the decision analysis in connection with loss of human lives and/or specify different types of probability bounds as for example bounds on related risk profiles. These criteria based on the Life Quality Index (LQI) proposed by Pandey (1997) and improved by (Skjong (2002) provide cost acceptability requirement is shown in Figure 2.5.

## 2.7.5  Risk Cost Benefit Analysis (RCBA)

This involve tool for organizing information on the relative value of alternative public investments and environmental impact. The value of all significant benefits and costs can be expressed in monetary terms. The net value (benefits minus costs) of the alternatives under consideration can be computed and used to identify the alternative that yields the greatest increase in public welfare. The economic benefit and risk reduction ascribed to each risk control options should be based on the event trees developed during the risk analysis. Estimates on expected downtime and repair costs in case of accidents should be based on statistics available (TETNR, 2006). Risk acceptability is the difference between the expected income and the expected operational cost. The net gain (g) from which all the running costs are subtracted. All expected losses (i.e. the risk) that may be caused by the occurrence of unwanted and unplanned events such as damages during storms, collisions, grounding, and fire are subtracted. Omitting any interest rates decision criteria can be represented by the following equation:

$$g - \sum_{i=1}^{N} \lambda_i \, \mu_o > 0$$

Eq. 2.8

N is the number of considered unwanted events, $\lambda_i$ is the frequency of occurrence of unwanted event category i; and $\mu_{oi}$ is expected loss of the system and the index o for owner, following the occurrence of unwanted event category. The term in the summation is the risk. Insurance risk premium can be paid to cover parts of any potential losses. Expected annual net gain minus the expected annual losses is large than zero, otherwise it will lead to bankruptcy. It is important to consider added value to society at the goal based objective formulation. If the society perceives a positive net benefit, the activity may be claimed to be corporate socially responsible. Incorporating this we may write (Nathwani et al., 1997).

$$\left( g - \sum_{i=1}^{N} \lambda_i \, \mu_o \right) r_p > \sum_{i=1}^{N} \lambda_i \cdot \max\{\mu_p - \mu_o \, ; 0\}$$

Eq. 2.9

$r_p$ a (political) factor is defined by society that specifies how much of the owner proceeds (yield); $\mu_{pi}$ is the expected loss to the society, in excess of owner's compensation, following the occurrence of unwanted event category i; Note index p is used for public (society). There are several indices that express the effectiveness of an RCO but currently only one is being extensively used in FSA applications. The Cost of Averting a Fatality (CAF) and can be expressed in the forms of Gross and Net. Gross Cost of Averting a Fatality (GCAF) = C/R.

## 2.7.6 Safety Management System

According to Bercha, (2004), managerial and organizational factors in accident causation are important in implementing outcome of risk work. The importance elements of safety management system include human reliability, safety assessment, and organisation structure. Eemergency preparedness and techniques of accounting for the safety management in risk based model have widely used in offshore and nuclear industry. Risk management system mostly depends on Judgmental approach, generic approach, suitable audit technique, human factors analysis, and global frequency modification (Wang, 2000). Risk management involves cost effective alternative risk reduction measures. Zero discharge should represent zero risk, but the challenge is to bring the risk to acceptable level and at the same time, derive the maximum benefit. Both pressures and measures should be identified though a common understanding. This goal should be achieved by an interdisciplinary process. Opportunities to improve both the environmental and navigation conditions through a joint approach need to be identified. Contemporary risk assessments involve evaluation of affected, monetary loss, equipment downtime, area affected, nature of credible accident and probability of adverse consequence hazard. Systematic process to quantitatively measure perceived risks values of waterways using input from waterway users and experts compensate for elements of sustainability under community participation is important.

## 2.7.7 Risk and Analysis Considerations

In addition to sound process, robust risk framework and eventual deductive risk model, there are other considerations that should be factored into the design of an effective risk and reliability model. These items could include the use and availability of data, the need to address human factors, areas of interest, and approaches to treating uncertainty in risk analysis. Data required for the risk work involves information on traffic patterns, the environment, historical, current operational performance data, and human performance data. As for mechanical failure they are independent of location. Mechanical failures often depend on factors like duty cycles or maintenance procedures, which, in turn, depend on the particular service in which the vessel is employed (Fotland, 2004). The model intention should reflect appropriate selected databases that accurately represent the local situation and the effectiveness of the models. However, there is always issue of missing data or data limitations in complex system analysis. Therefore, procedures are required to develop compensation for such situation in relation to the system. The model could use probabilistic, simulation, stochastic and expert judgments couple with deterministic method to cover such situation for reliable system analysis and design. A safety culture questionnaire which assesses organizational, vessel safety and climate can be administered to provide qualitative and quantitative input to the analysis (Akita, 1972).

### 2.7.7.1 IWT Safety Risk Factors Consideration

In order to improve safety in waterways, risks that need to be avoided and taken into consideration when necessary in the model developments are A i) Hazardous substances on board ships, ii) shipboard hazard to personnel, iii) oil in galley iv) equipment, v) machinery spaces, vi) potential source of ignition, viii) deck areas. B) Waterways hazards involve risk from storms, lightning, uncharted submerged objects, other ships, flood risk, collision risk, stranding, and grounding. C) HAZID for ship motion in heavy weather is also considered important contributing factors in the analysis of specific risk scenarios. A number of accident associated with heavy weather that may be incorporated into risk model are water ingress, parametric rolling. D) Technical, operational, and organizational risk reduction measures may be identified and evaluated by the use of a risk model. Other operational hazards like cargo lashing failure, propulsion failure, mooring failure and lifting failure could also be independently analyzed (Bottelberghs, 1995).

### 2.7.7.2 IWT Environmental Risk Factors Consideration

Occurrence of accident within the coastline is quite prohibitive due to unimaginable consequences and effect to coastal habitat. Recent research on transportation mode has shown that IWT provide wide advantages that translate to potential need for its use in near future See Table 2.10 (Hennigsen, 2000).

**Table 2.10:** Emission transportation mode comparison

| Mode | Operation Energy | Hauled Energy | Modal Energy |
|------|------------------|---------------|--------------|
| Rail | 412.5 | 706.3 | 1075 |
| Truck | 1312.5 | 1312.5 | 2137.5 |
| Barge | 262.5 | 262.5 | 618.8 |

A group of expert on the scientific aspects of marine pollution on the state of marine environment, 1989 stated that: "Man's fingerprint is found everywhere in the oceans. Chemical contamination and litter can be observed from the poles to the tropics and from beaches to abyssal depths. But conditions in the marine environment vary widely. The open sea is still relatively clean. In contrast to the open ocean, the margins of the sea are affected by man almost everywhere, and encroachment on coastal areas continues worldwide. If unchecked, this trend will lead to global deterioration in the quality and productivity of the marine environment".

The above quotation describe the extent and various ways to which human activities and uses water resources affect the ecological and chemical status of large river systems. From ecological point of view, navigation is not the only human pressure derived activities. Activities like hydroelectric power, river straightening and flood control play significant role in river engineering and hydro morphological situation. Sustainable solutions for river and IWT require an integral perspective and multidisciplinary answers. IWT potential and environmental benefit could be quantified into benefit in the risk and reliability process. Design based on risk and reliability can augment its environmental and safety performance. The significant of sustainable balancing of economic, environmental development, community involvement maximize benefits of the planning and implementation strategy. Such planning could result to dramatically improved public access, provision of new open spaces, improved quality of life, strengthened city, image community approval and pride. The above benefit could be quantified into risk cost benefit analysis in the sustainable risk and reliability process.

### 2.7.7.3 Channel Improvement Consideration

Channel improvement include dredging, VTS, AIS, alignment. According to their goal, dredging activities can be classified into capital, maintenance and remedial works. Rivers carry suspended sand and soil along with them as they flow toward the ocean. The higher the water velocity, the greater its energy and capacity to move soil, sand, and even rocks along with it. Sediments can build up and become shoals, or high spots in the navigation channel. The consequence of this could lead to heavy vessel or infrastructure damage or even environment damage if the ship's cargo if spilled into the water. This makes it imperative to keep the channels safe for navigation. Prior to dredging, hydrographic work is necessary to provide update on the condition of the water, record keeping on ship size and behavior or response.

When there is a significant area of water that is too shallow, dredging should be carried out. Bathymetric mapping is very complicated. It should shows that all depth measurements are relative to the surface of the water. Measurements should be adjusted to compensate for tidal and seasonal variations in water depth measurements. In water bodies not affected by tidal variations, the depth is referenced to another datum and the surface of the water is adjusted to other constant datum, mean lower low water (MLLW) in coastal areas (Stamatis, 2004).

Methods for estimating total volumes of dredge include three general methods are average end area method; borrow pit method, triangle method. Estimates of rock excavation and common excavation are needed with the higher degree of accuracy when both soil and rock are encountered at depths above allowable over depth. Dredging equipment play a very important role in maintenance of channel, and selection of best dredger for sustainability of channel. There are various types of dredging equipment, with different characteristics.

### 2.7.7.4 Integrative Components of IWT

Earlier technological development aimed at better using facilities provided by nature which required less energy and produces less waste. Modern technology tends to replace these by new ones that include natural soil fertility by chemicals fertilizers, inland waterway by roads and railways, hydropower by chemical power (coal, oil or nuclear). This is required mitigate environmental revolt of today.

### 2.7.5 Human Reliability Assessment

The main causes of accidents is shown in Figure 2.8, where the first 60% of the total number of claims recorded that human error as the direct cause and further 30% human error is from indirect contributory cause (IMO, STCW, 95).

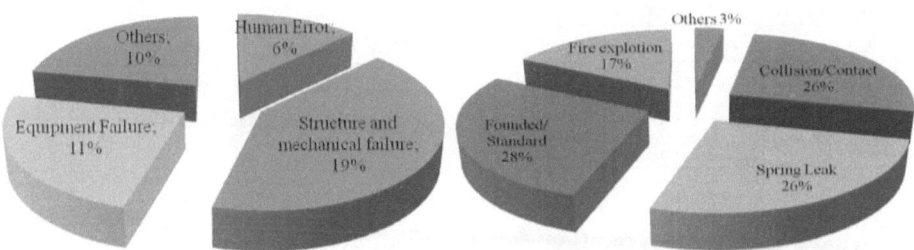

**Figure 2.8:** Main cause of accidents (Marine Department, 2009)

Assessing the role of human and organizational performance on levels of risk in the system is important, especially when such error is often cited as a primary contributor to accidents (ILO, 1976). Human factors that trigger of human errors are the main source of risk in maritime activities. Modest approach that allow evaluating quantitatively and qualitatively of the real incidence of several human factors over maritime accidents happening should be utilized for the assessment and analysis. The process should be with the objective of eliminating human factors while taking into consideration properly developing risk management plan (Catherine et al, 2006).

## 2.7.6 Validation and Reliability

Validation work is required to cover the whole system, multiple system, complicated subsystems and components. Each of which must be modeled and validated. The current state of practice often attempts to validate system models directly from test data taken on the entire system. This approach can be problematic if there are a large number of components or if subsystem models contain complex connections or interfaces, energy dissipation mechanisms, or highly nonlinear behavior. There is hierarchy of system on which model can be built and validation carried out on. The top tier is the complete system (Skjong et al., 2005). Table 2.11 show the best practice model validation works.

**Table 2.11:** Best practice of model verification and validation

| Institution | Model |
|---|---|
| ASME | Committee for development of validation procedure for computational and solid mechanics |
| DOE | Science based (numerical & simulation) stockpile stewardship program for high level safety, reliability & performance of nuclear stockpile |
| Los Alamo National Lab | Reliable, numerical & predictive method for behavior of material under high pressure and material |
| AIAA | Computational fluid dynamics V&V committee work |
| NASA | The development of guidelines and recommended practice for model V&V to provide a foundation for increased use of numerical simulation for certification |

Model validation is required when a predictive model is the end product. A model is the conceptual, mathematical and numerical description of a specific physical scenario. This includes geometrical, material, initial, and boundary data. Backdrops associated with validation include historical data limitation, problem associated with complex system, inability to perform controlled experiments on the system, lack of real world system, agent behavior, and interaction mechanism, emergent and random system. Pathway to validation encompasses exhaustive exploration of critical cases. Practical validation requires iteration work to put the model through active nonlinear tests in order to explicitly formulate a series of mathematical tests. The end result of validation can leads to a better understanding of the model's capabilities, limitations and appropriateness for a valid conclusion and addressing a range of important questions.

Other mathematical theories such as fuzzy sets, evidence theory, the theory of random sets, and the theory of information gap can also be used for validation of prediction model. Thus, their application is less efficient compare to probabilistic methods. When the standard deviation in the model input parameters has been established, this variability can be propagated through the simulation to establish an expected deviation on the model output quantities. Sampling based propagation methods Monte Carlo, Latin Hypercube are commonly used, where samples are drawn from the input parameter populations, evaluate through deterministic model, and plot distribution of the response quantities. Sensitivity based methods such as First Order Reliability Methods (FORM), Advanced Mean Value (AMV), and Adaptive Importance Sampling (AIS) are more efficient than sampling based methods. Sampling based methods is used to propagate input uncertainties to uncertainties on the response quantities.

## 2.7.7 System Uncertainty Analysis

Uncertainty is part of risk, but its abstract nature and limitation of knowledge of unseen in real world settings make its quantification a complex work. Associated with uncertainty are normally reflect issue of influences. This include recovery process, test of new advancements, influence on policy, address of system changes over time, services and resources. Science enters into decision making processes in local public administrations in several ways. For example science is integrated within the administration through a body of civil servants with position as technical expertise. One informal way in which science may influence policy is through creating public awareness on certain issues, choosing the issues that should be on the agenda. Simulation also offers one best option to cover extreme case uncertainty beside probability. Two common uncertainties as described by Emi et al., (1997) are aleatory uncertainty and epistemic uncertainty. Aleatory uncertainty is represented by probability models that give probabilistic risk analysis its name, while epistemic uncertainty is represented by lack of knowledge concerning the parameters of the model. In the same manner of addressing aleatory, epistemic uncertainty can be addressed through probabilistic risk analysis.

To deal with uncertainty it is important to consider qualification of knowledge base, the values and biases in the various choices and assumptions involved. There are three dimensions of uncertainty: The location of uncertainty (where within the model). The level of uncertainty (from deterministic knowledge to total ignorance); and the nature of uncertainty (whether the uncertainty is due to the imperfection of our knowledge or is due to the inherent variability of the phenomena being described).

## 2.7.8 IWT Regulation

Due to international implication of maritime industry, the required laws to be implemented are finalized by UN agencies following tacit procedure. The states decide on formulating local legislation towards implementation through marine administration, port and flag state control. Three main purposes of legislation under legal framework are: To provide legal framework for maritime transportation, for implementation basic objectives of a states, and to achievement of certain economic purpose. Because of the issues of environment problem, many other environmental directives, policies and conventions interface are being created according to needs. Such a standard could also be used to improve the safety of navigation and protection of the marine environment (IMO, 2006).

In reality ships spend 90 percent of their operational lives at high speed in deep sea water. It is during the mandatory beginning and end of every voyage when the risk of collisions and groundings are highest because of proximity to port. Ensuring the ability to maintain complete and positive control of a ship's movement during these segments of a voyage is absolutely vital if that risk is to be reduced. There is no recognized measure or point at which a channel is identifiably substandard. Channel improvements should ideally keep up with traffic so that a channel never becomes substandard. PIANC approach to channel design guide provides the basic assumptions drawn from information sharing in PIANC working group studies and symposium. They contain many significant articles addressing issues ranging from technical, maintenance to policy and regulatory on aspects of navigability constraint for waterways (PIANC, 2002).

## 2.7.9 Sustainability

Sustainability emphasize on efficiency, and life span of the system, it deals with proactive approach for assurance of environment protection, survival, reliability and preservation of the planet. Prevention of accident through safety and cost effective decision on system requirement contribute to construct of barrier that prevent accident to happen in the first place pollution that precede such accident not to happen. The prevention of maritime industry need to adjust to the ways we do things in a world of sensitivity being characterized by sustainability, capacity building, efficiency, optimization of development, practice and operations that meets the needs of the present generation without compromising the ability of future generation to meet their need. For channel work, sustainability focuses on For operators, larger traffic in limited depth waterways, reconciliation between safety and complex efficiency challenge, both to the regulatory and operational agencies, ii) The regulatory agencies, it is extremely important to ensure that safety is not compromised for the sake of efficiency, and iii) Operational agencies, it is equally important that efficiency is not compromised.

An environmental issue has difficulties associated with changes to the bathymetry due to channel work or as a resulted in changes in water currents or other oceanographic effects. Or as result of sediment transport and need to maintain, sustain our living, existence and purpose associated with them, that require historical as well as recent and predictive datasets system. Predictions of these parameters with the use of numerical calculation models that can provides real time information about natural system behavior of water levels, currents, other oceanographic, meteorological data from bays and harbors, are important..

Most IWT has been under utilized and under maintained especially for transportation. The Butts et al., (1992) studies outcome promise potential for rise. IWT, this include logistics for short sea and supply vessel operation to accommodate requirement of growing vessel size and expanding deep sea operation, also the use of high speed craft for recreation, and the use of barges for freight transportation. IWT development represents complex and dynamic system that require be designing and establishing based on concrete proactive approach.

Inland water cannot stand alone, both IWT and ecological integrity have certain basic needs. In order to develop mutually acceptable solutions, such needs must first be clearly defined. The implementation of a new, integrated planning philosophy under risk and goal based design would help ensure both sustainable development of IWT and the achievement of all required environmental objectives. Efforts is required to reduce the impact on aquatic ecological integrity including non structural measures to improve inland water navigation components like fleet development, new ship technology, inter modal connections, river information systems. Defining this level required the interpretation of the case, and implied to give a normative resolutions task to the system. The use of hybrid of system and scientific based, knowledge and social problems is also recommended.

# 3

## Research Methodology

### 3.1 Research Methodology Flowchart

Qualitatively and quantitatively approach is employed to carry out this study in deduce the risk of collision, enhance design against accident load in IWT and propose mitigation from the analysis. The Techniques employed include HAZID, frequency and consequence risk analysis, Failure Modes and Effects Analysis (FMEA), Fault Tree Analysis (FTA) and Event Tree Analysis (ETA), Cost Effectiveness analysis and Risk control option. The flow chart in Figure 3.1 shows the process involved in this study this study.

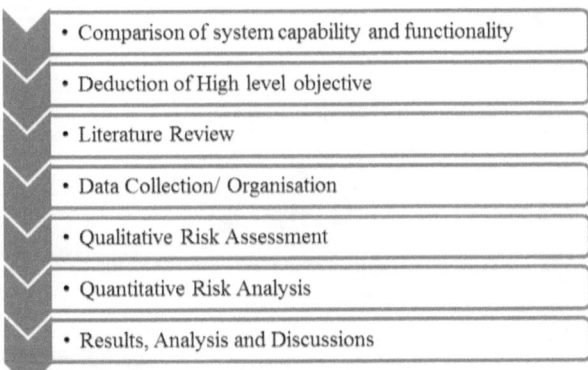

**Figure 3.1:** Research Methodology flowchart

## 3.2 Identification of Research Variables

### 3.2.1 Safety Risk and Reliability Model

This chapter discusses the best practice approach for the development of safety risk and reliability model for IWT, protection of the marine environmental which is a proponent of consequence of collision where assessment, evaluation of risk and validation with goal base life cycle requirement. Elements that will optimize design, existing practice, and innovative entity and facilitate decision support for policy accommodation for evolving coastal transportation regime is also discussed in this chapter. liability based verification and validation of system in risk analysis should be followed with creation of database and identification of novel technologies required for implementation of sustainable system. Risk framework has been described in Chapter 1, detail of the model and process is described in this chapter. This chapter presents the modeling of safety and risk and reliability model (SARM) for sustainable IWT. Effective risk assessments and analysis required three elements highlighted in the relation below.

Risk modeling = Framework + Models + Process                Eq. 3.1

Figure 3.2 shows the flowchart of safety and reliability model flowchart.

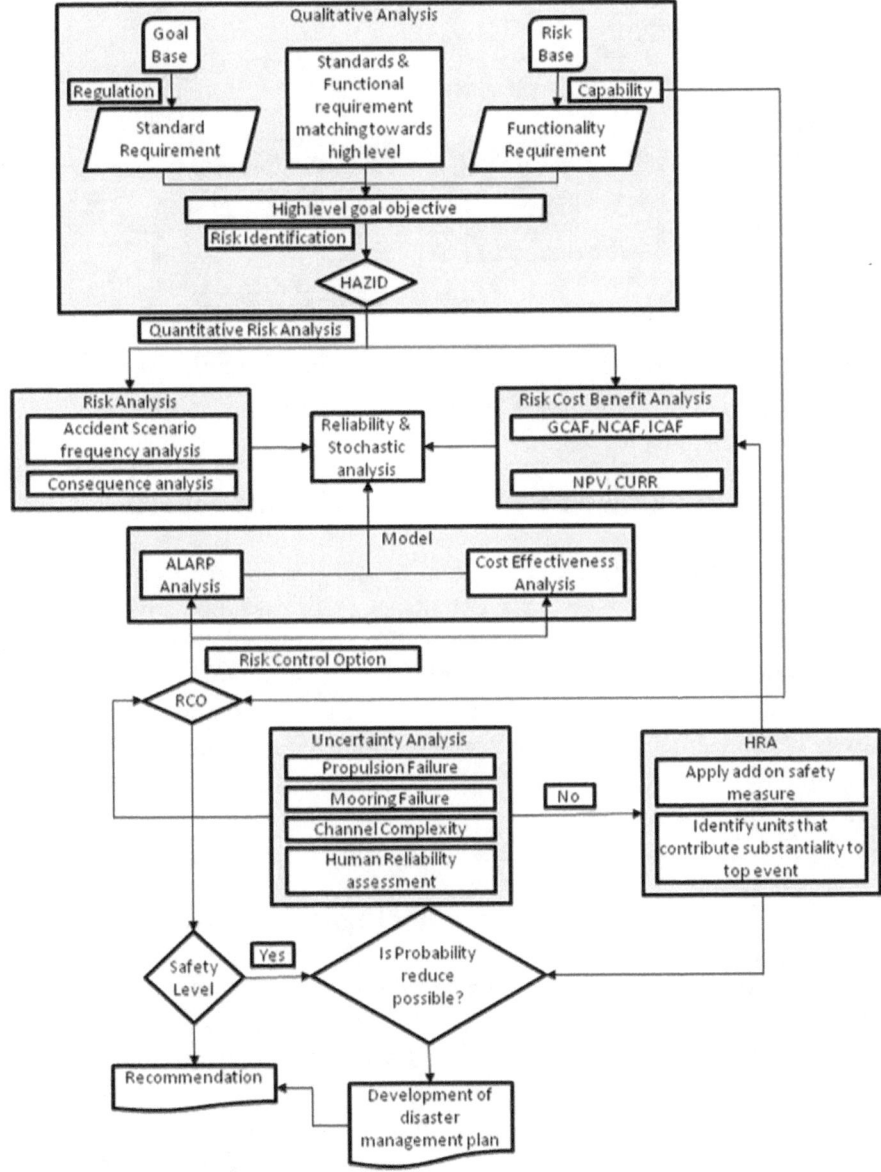

**Figure 3.2**: Safety Risk and Reliability Model flowcharts

## 3.2.2 Safety Risk and Reliability Model Framework

Figure 3.3 shows the flowchart of safety and reliability model framework.

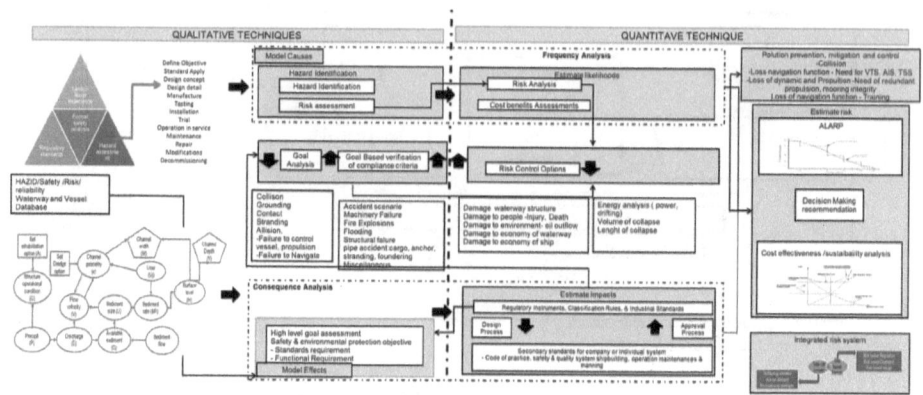

**Figure 3.3**: SARM framework

## 3.2.3 Elements of the Safety Risk and Reliability Model

According IMO (2002), conceptual model presented in chapter 1, risk analysis involves the following elements i) goal base objective ii) hazard identification, risk assessment and criticality ranking, iii) frequency analysis and consequence risk analysis iv) evaluation of cost effectiveness analysis and risk control option and v) Decision and Risk acceptability criteria. The analysis of the elements can be performed under the following category

i.    Qualitative Analysis.

   a.    Element 1: Goal based Objective
   b.    Element 2: Risk Assessment: HAZID operational identification involve understand situation background (consider each hazard; identify all potential hazardous scenarios which could lead to significant consequences, HAZARD (Risk Ranking and Matrix development) including classifying probability, classifying consequences. Risk matrix focus efforts on the serious ones and normally has three regions (Negligible, Tolerable and Intolerable). Various approach could be used to get this done, this include the use of HAZOP and PrHA and Failure mode (FMEA), checklist, however, combination of approach is recommended.

The link of HAZID is established using the risk matrix. The risk matrix can be constructed as shown in Table 3.11. Risk Matrices are the most important tools that are provided to the group of experts and are being used to accomplish the previously mentioned task of ranking. Low—consequence events over extremely rare accidents that is really catastrophic. A better type or risk matrix should be defined that should also lend itself to environmental protection issues. Two-dimensional approach could be adopted where, one that retains both dimensions of risk instead of combining them into a single number. Even so, a scheme for the ranking of different (frequency-severity) combinations should be devised, something that would necessitate a more systematic investigation whether the decision-maker is risk averse, risk neutral, or risk prone. Risk acceptability of one fatality is somehow equivalent to 10 severe injuries, something that can be debated at least on ethical grounds. Risk index is developed following equation 3.2.

Risk Index = Frequency Index + Severity Index          Eq. 3.2

ii.  Quantitative, Probabilistic, Stochastic and Reliability

   a.  Element 3: Risk Analysis: Frequency estimation analysis, Consequence estimation analysis.
   b.  Element 4: Risk control option (RCO) and risk control measure. RCO is used to propose effective and practical Risk Control Options (RCOs) comprising the following four principal stages focusing on risk areas needing control; identifying potential risk control measures (RCMs); evaluating the effectiveness of the RCMs in reducing risk by re-evaluating iterating the system and grouping RCMs into practical regulatory options. Risk Control Measures, through expert meetings, are combined into potential Risk Control Options.
   c.  Element 5: Cost effectiveness analysis (CEA), Risk, Cost benefit analysis, safety level analysis, and environmental tolerance analysis is a quantitative approach to estimate and compare the cost effectiveness of each option in terms of the cost per unit risk reduction. The cost component consists of the one-time (initial) and running costs of an RCO, cumulating over the lifetime of the system. The benefit part is much more intricate. Cost is usually expressed using monetary units. Consequences that deal with fatalities are considered in this step, although attempts to extend it to environmental consequences, decision and risk development of risk management system, monitoring, implementation, supervision of control and uncertainty are also considered.

d.  Element 6: Recommendation for decision making and risk acceptability criteria; provision of recommendation for decision for safety improvement. The RCOs that are being recommended should reduce Risk to the desired level and be cost effective.

e.  Risk acceptability criteria (RAC). Acceptance criteria are established to impose restrictions on the engineering designs and operations to keep the consequences of adverse events below some limits. Some restrictions amount to a loss of benefit both to the society and to the individual owners. They represent rational principals established to support optimization of designs and operations. Acceptance criteria may be deterministic, probabilistic, or semi-probabilistic that indicates acceptable safety, limits to the probability of occurrence of adverse events, or boundary on the probability of consequences.

    −   Desired Risk Level: Individual and societal types of risk should be considered for crew members, passengers and third parties. Individual Risk can be regarded as the risk to an individual in isolation while Societal Risk as the risk to the society of a major accident.

    −   As Low As Reasonably Practicable (ALARP): According to Health and Safety Executive's (HSE, 2001) Framework for the tolerance of risk, there are three regions in which risk can fall into Unacceptable Risk for the case resulting from high accident frequency and high number of fatalities should either be forbidden or reduced at any cost. Between this region and the Acceptable Risk region, ALARP region is defined. Risk that is falling in this region should be reduced until it is no longer reasonable to reduce the risk, Acceptance of an activity whose risk falls in the ALARP region depends on cost-benefit analysis.

    −   Cost-Effectiveness Criteria: As mentioned before, acceptance of shipping activity whose risk falls in the ALARP region depends on cost-benefit analysis. Also important is cost-effectiveness indices and the "$3m criterion". The proposed values for NCAF and GCAF in Table 3.1 have been derived by considering societal indicators (refer to MSC 72/16, UNDP 1990. The value of $3 million is based on the Implied Cost of Averting a Fatality (ICAF) and has been calculated using OECD data.

f. Individual Risk Acceptance Criteria: IMO's guidelines provide no Risk Acceptance Criteria; currently decisions are based on those published by the United Kingdom Health & Safety Executive (HSE, 2001). The IMO adopted HSE's criteria that define the intolerable and the negligible risk for a single fatality. Risks below the tolerable level but above the negligible risk (for crew members, passengers and third parties) should be made ALARP by adopting cost-effective RCOs (Table 3.2).

g. Societal Risk Acceptance Criteria: The purpose of societal risk acceptance criteria is to limit the risks from ships to society as a whole, and to local communities (such as ports) which may be affected by ship activities. In particular, societal risk acceptance criteria are used to limit the risks of catastrophes affecting many people at the same time, since society is concerned about such events.

h. Acceptance criteria and performance of a ship in an accident can be measured by kinetic energy dissipated in damaged structures, penetration depth in an accident, quantity of oil outflow, and/or and residual hull girder strength. The acceptance criteria include minimum distance of cargo containment from the shell; ship speed above which a critical event happens; allowable quantity of oil outflow; and/or minimum values of section modulus or ultimate strength of hull. A design should demonstrate that cargo tanks/holds are not breached in an accident so that there will be no danger of pollution or, if the cargo tanks are breached, the oil out-flows following an accident is limited; and the ship has adequate residual hull girder strength.

**Table 3.1:** Cost Effectiveness Criteria

|  | GCAF | NCAF |
|---|---|---|
| Criterion for Risk of Injury or ill health | 3 Million | |
| Criterion convering only | 1.5 Million | |
| Criterion convering only risk of injury and ill health | 1.5 Million | |

**Table 3.2:** Tolerable level

| Risk Level | Acceptability |
|---|---|
| Maximum tolerable risk for crew members | 1.0E-02 |
| Maximum tolerable risk for passengers | 1.0E-03 |
| Maximum tolerable risk for public ashore | 1.0E-03 |
| Negligible risk | 1.0E-05 |

## 3.3 SARM Model Components Development

There is various risk and reliability tools available for risk based methods that fall under quantitative and qualitative analysis. Choice of best methods for reliability objective depends on data availability, system type and purpose. However employment of hybrid of methods of selected tool can always give the best of what is expect of system reliability and reduced risk. Figure 3.4 shows the risk and reliability model combined process diagram. The analysis is a purely technical risk analysis. When the frequencies and consequences of each modeled event have been estimated, they can be combined to form measures of overall risk including damage, loss of life or propulsion, oil spill. Various forms of risk presentation may be used. Risk to life is often expressed in two complementary forms. The risk experienced by an individual person and societal risk. The risk experienced by the whole group of people exposed to the hazard and damage or oil spill.

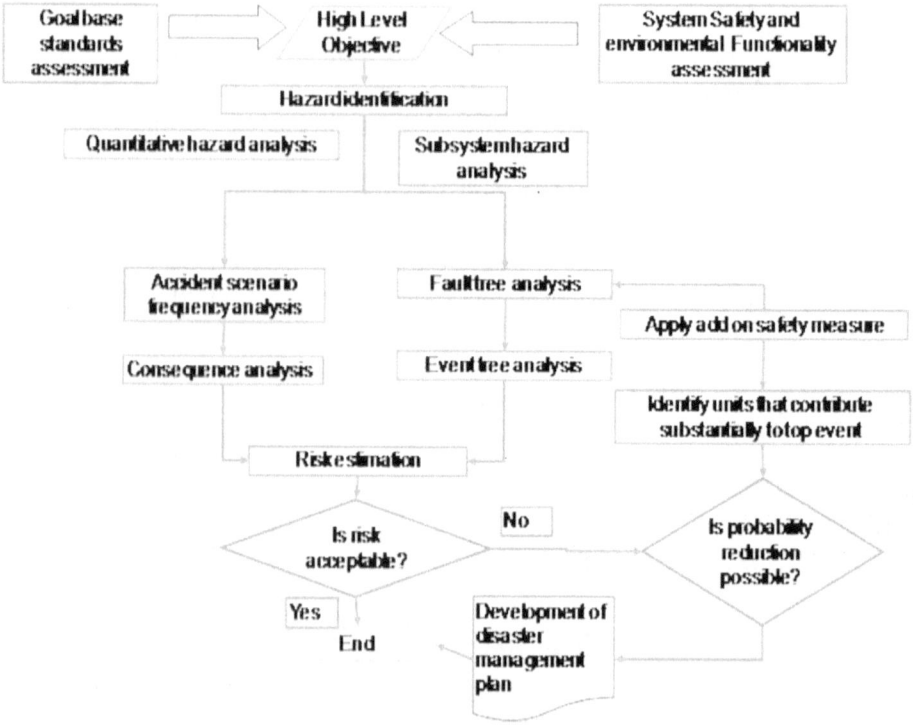

**Figure 3.4**: SARM High level Objective

## 3.3.1 SARM Process

SARM intend to address risks over the entire life of the complex system like IWT system where the risks are high or the potential for risk reduction is greatest. SARM address quantitatively, accident frequency and consequence of IWT. Other risk and reliability components including human reliability assessment, it is recommended to be carried out separately as integral part of integrated risk process. Other waterways and vessel requirement factors that are considered in SERM model are construction, towing operations and abandonment of ship, installation, hook-up and commissioning and development and major modifications. Integrated risk based method combined various technique as required in a process. Table 3.3 shows risk based design for techniques. This is applied for each level of risk. Each level is complimented by applying causal analysis, expert analysis, and organizational analysis in the risk process.

**Table 3.3**: Risk based design techniques

| Process | Suitable techniques |
|---|---|
| HAZID | HAZOP, What if analysis, FMEA, FMECA |
| Risk analysis | Frequency, consequence, FTA, ETA |
| Risk evaluation | Influence diagram, decision analysis |
| Risk control option | Regulatory, economic, environmental, function elements matching and iteration |
| Cost benefit analysis | ICAF, Net Benefit |
| Human reliability | Simulation/ Probabilistic |
| Uncertainty | Simulation/probabilistic |
| Risk Management/ monitoring | Simulation/ probabilistic |

Figure 3.4 shows stakeholder that are considered in risk process. The stakeholder is taken into consideration at the decision phase, weather decision that will be adopted will satisfy them or not. Technically, the process of risk and reliability study involves the following four parts:

i. System definition of high goal objective: This requires defining the waterways by capturing gap between system functionality and standards. This involves definition and problem identification: functionality and standards requirement, gap identification and setting high level objective for risk prevention method.

ii. Qualitative hazard identification and assessment. It involves hazard identification through qualitative review and assessment of possible accidents that may occur, based on previous accident as well as experience or judgment of system users where necessary. It also involves hazard and consequence identification: risk ranking and setting acceptance standards for the risks.

iii. Quantitative hazard frequency and consequence analysis: once the hazards have been identified and assessed qualitatively. Frequency analysis involves estimation of how likely it is for the accidents to occur. Analyzing the likelihood's of occurrence: and risks of possible events, normally express in rate of accident per year. I also involve analyzing consequences: Impact, loss of life, damage to infrastructure, economic loss, penalty, injury, oil spill, environmental degradation.

iv. Evaluation of cost effectiveness and risk control option, this involve evaluation of uncertainty, and risk control measure (RCM)

v. Risk acceptability and sustainability analysis; the yardsticks to indicate whether the risks are acceptable, in order to make some other judgment about their significance. This involves checking of non-technical issues of risk acceptability and decision making as well as reliability based model verification and validation. This step falls under concept of risk management, it has component of decision making tool for risk acceptability and reduction measures but it still need to be carried out. Figure 3.5 shows SARM process diagram.

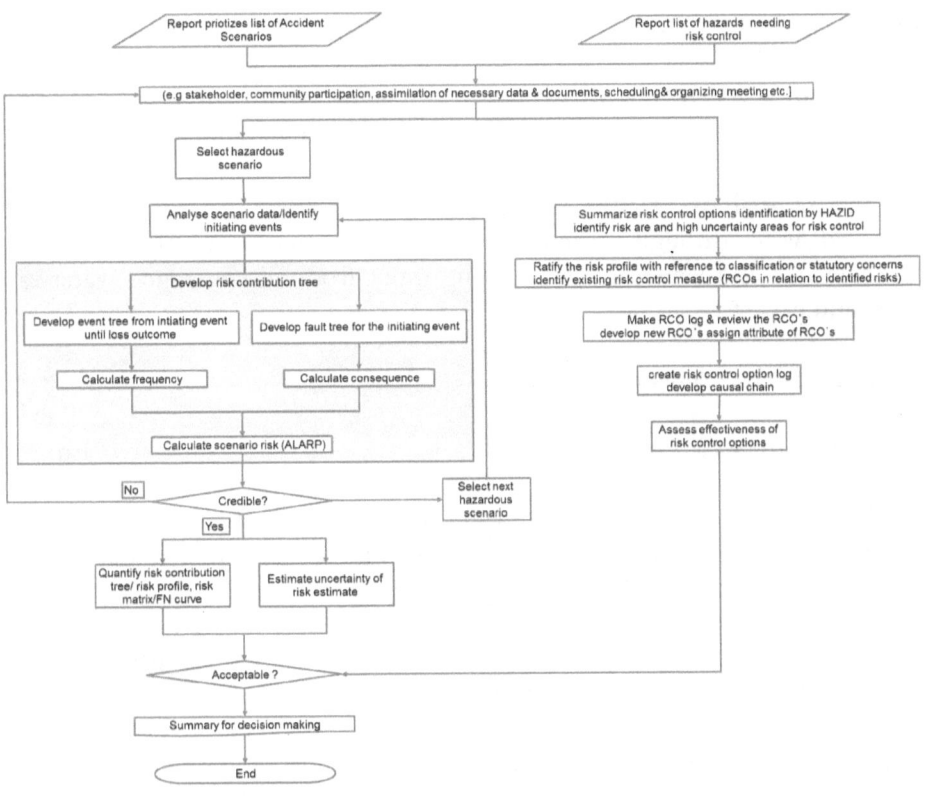

**Figure 3.5:** Risk and reliability process flowchart

Proactive risk and reliability based design is beneficial to achieve the following in system engineering of IWT: i) Estimating risk levels and assessing their significance. ii) Identifying the main contributors to the risk helps in understanding of the nature of the hazards and suggests possible targets for risk reduction measures. iii) Defining design accident scenario iv) Comparing design options gives input on risk issues for the selection of a concept, v) Evaluating risk reduction measures though sustainability analysis through cost benefit, safety and environmental cost translation towards a cost-effective risk reduction solution, vi) Demonstrating acceptability to regulators and the workforce for beyond compliance vii) Identifying safety critical procedures and equipment that can provide minimize risks operability, process like giving provide attention and look out procedure during operation. viii) Identifying accident precursors which may be monitored during operation to provide trends of incidents and preparedness for solution. Risk defined as:

Risk = Probability (Pa) x Consequence (Ca)                      Eq. 3.3

Or in a more elaborate expression risk can be defined as:

Risk = Threat x Vulnerability x
{Direct (short-term) consequences + (broad) Consequences}   Eq. 3.4

In risk analysis, serenity and probability of adverse consequence hazard are deal with through systematic process that quantitatively measure , perceive risk and value of system using input from all concerned waterway users and experts.

Risk can also be expressed as:

Risk = Hazard x Exposure                                      Eq. 3.5

Where hazard is anything that can cause harm (e.g. chemicals, electricity, Natural disasters), while exposure is an estimate on probability that certain toxicity will be realized Severity may be measured by no. of people affected, monetary loss, equipment downtime and area affected by nature of credible accident. Risk management is the evaluation of alternative risk reduction measures and the implementation of those that appear cost effective where:

Zero discharge or negative damage = Zero risk                 Eq. 3.6

## 3.3.2 Risk Analytical Method for Inland Waterway Components and Data Requirement

Table 3.4 shows typical phases of risk components in IWT project phase.

**Table 3.4:** IWT project phases risk components

| Project Phase | Planning and design | Construction | Post construction |
|---------------|---------------------|--------------|-------------------|
| Impact | Need for IWT is translated into project design Physical and biological impacts depend on project specification. | Project construction may cause temporary or permanents physical change, Advert effect should be mitigated through best practice method | Physical change may result to long term environmental effects that should be mitigated by appropriate project design , planning and execution |

| Scope | Functional requirements Conceptual design Potential environmental impacts, Final design and specifications | Tendering and contract award Construction methods and equipment' selection Monitoring and feedback | Infrastructures in service and there may have additional mode of impact Long term monitoring and feedback. |
|---|---|---|---|
| Environmental components | Planning and design decision RCO to prevent or reduce environmental impact if of the whole project | Construction decision RCO to prevent environment impact cause by physical change | Certain RCO may apply to mitigation of future impacts |

## 3.4 SARM Model for IWT Functionality and Standard

Risk and reliability based model aims to develop innovative methods and tools to assess operational, accidental and catastrophic scenarios. It requires accounting for the human element, and integrates them as required into the design environment. To pursue this activity effectively, an integrated design environment to facilitate and support a holistic risk approach to ship and channel system. Figure 3.6 shows possible accident scenario in waterway. The following are components are included in the process:

i. The earlier stage of the process involves finding the cause of risk, level of impact, destination and putting a barrier by all mean in the pathway.

ii. Secondly, cause of risk and risk assessment, IWT risk can be as a result of their root cause: Inadequate operator knowledge, skills or abilities, or the lack of a safety management system in an organization. Immediate cause: failure to apply basic knowledge, skills, or abilities, or an operator impaired by drugs or alcohol. Situation causal factor: Number of participants, time, planning, volatility environmental factors, congestion and time of day risk associated with the system. Organization causal factor: Organization type, regulatory environment, organizational experience management type, changes, system redundancy, system incident, accident, history, individual, team training and safety management system.

iii. Thirdly, risk work process targets are risk analysis and reduction process. Figure 3.7 shows the conceptual components of risk and reliability based design. The overlap represents the need for holistic approach consideration in risk analysis.

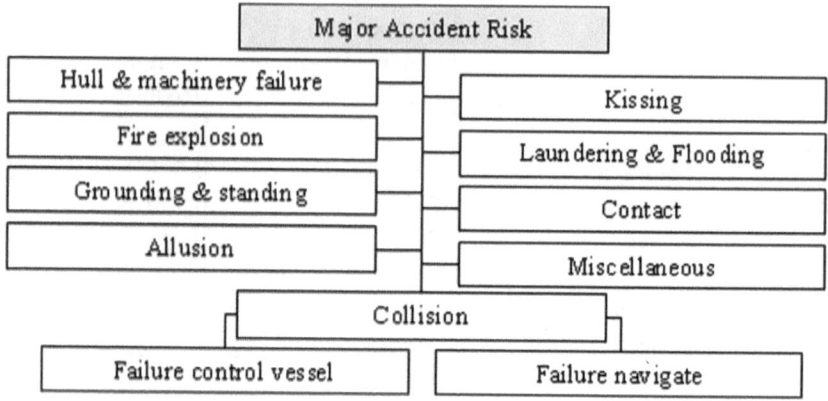

**Figure 3.6**: Major marine accident risk scenarios

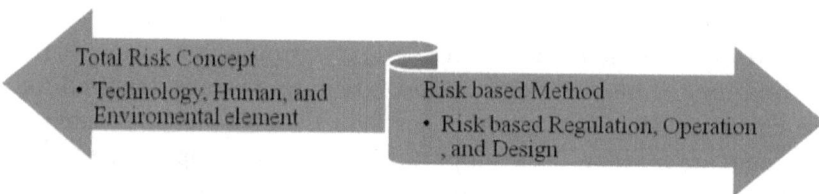

**Figure 3.7**: Risk based modeling concept

Uncertainty risk in complex systems can have its roots in a number of factors ranging from performance, new technology usage, human error as well as organizational cultures. They may support risk taking, or fail to sufficiently encourage risk aversion. To deal with difficulties of uncertainty risk migration in marine system dynamic, risk analysis models can be used to capture the system complex issues, as well as the patterns of risk migration. Historical analyses of system performance are important to establish system performance benchmarks that can identify patterns of triggering events, this may require long periods of time to develop and detect. Assessments of the role of human and organizational error, and its impact on levels of risk in the system, are critical in distributed, large scale dynamic systems like IWT couple with associated limited physical oversight. Marine transportation service provides substantial support to various human activities that has long been recognized. IMO long cross boundary maritime regulation activities has lesson learnt that could be modeled for today global regulatory quest for environmental challenges and advancement of human civilization. Most IMO regulatory work is not mandatory on coastal transportation. Except implementation of international transportation regulation that is directed through flag state control (FSC) state and port state control (PSC). Global accident database from IMO, MIAB and DnV, Norwegian waterway is used to assess accident risk and risk scenario on waterway and their associated seriousness. Norway Northsea and inland water represent the harshest water environment for marine transportation. Table 3.5 represent accident data in Norwegian waterway by Norwegian Maritime Directory (NMD) from 1997-2006.

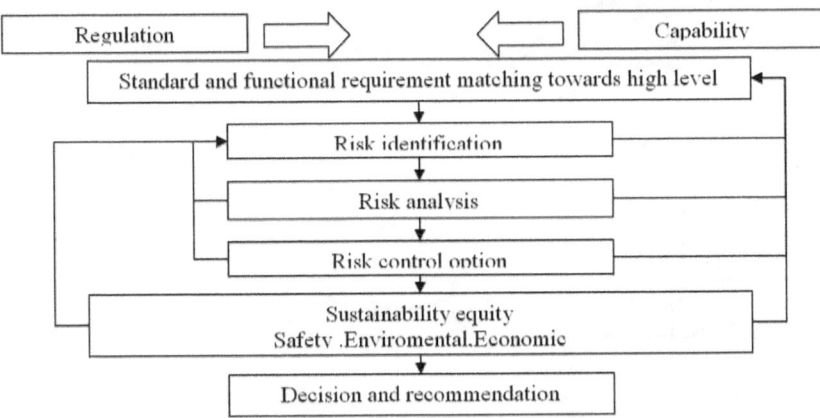

**Figure 3.8**: Risk and Reliability model Iteration flowcharts

**Table 3.5:** Number of fatalities or injuries by accident type (NMD, 2008)

| Cause | Fatality | Injury |
|---|---|---|
| Personal accident | 20 | 522 |
| Collision | 7 | 1 |
| Capsizing | 11 | 0 |
| Fire/Explosion | 0 | 3 |
| Others accident | 1 | |
| Total | 39 | 526 |

Figure 3.8 shows the flowchart of risk and reliability process iteration. Qualitative matching regulatory and functionality matching input to the system at the beginning of the study. Figure 3.9 shows components of vessel system assessment for ship and Figure 3.10 for waterway. This is followed by HAZID, in carrying out a hazard assessment it is vital that there are clearly defined objectives in terms of what is to be demonstrated.

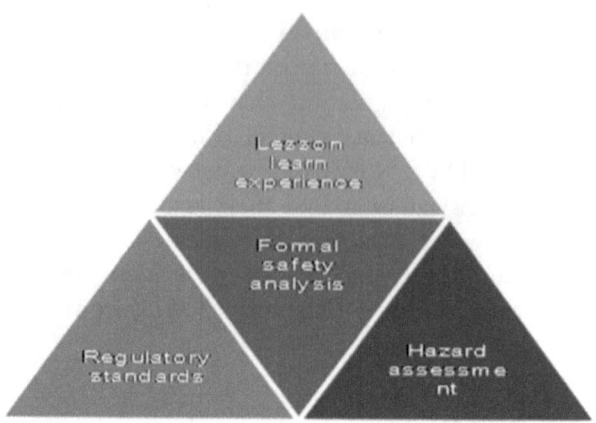

**Figure 3.9:** Ship risk components

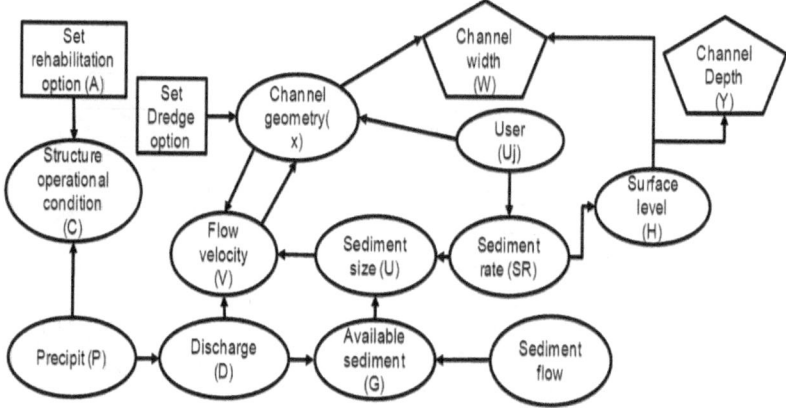

**Figure 3.10:** Waterway risk Components

## 3.4.1 Channel Parameters Consideration

Table 3.6 shows waterways width and depth requirement. The listing under each represent the component of data requires to be considering for category of risk analysis.

**Table 3.6:** Waterways width and depth requirements

| Width | Depth |
|---|---|
| Maneuvering lane | Draught |
| Bank suction | Trim |

| Wind effect | Squat |
| Current effect | Exposure allowance |
| Channel with bends | Freshwater adjustment |
| Navigationaids | Maneuvering allowance |
| Pilot | Overdepth allowance |
| Tugs | Depth transition, tidal allowance |

The risk technique involve preparatory work on problem definition that assess and match constrains requirement related to goals, systems, regulation and operations as well as depth and extend of the application and process. The preparation involves definition of deficient system operations, external influences, system category and to remedial recommendation for deficiencies and exclusion major risk categories.

### 3.4.2 IWT Vessels Parameter Consideration

Shipping is about port and access to port by optimum size of ships and its associated economics implication can be made available through navigable channels, ship design and choice of logistics. Ship production and condition of channels are likely to be out of phase. Economies of scale and demand have begot big ship to emerge within a short period of time after Second World War. Figure 3.11 shows the waterways width parameters and sub parameters.

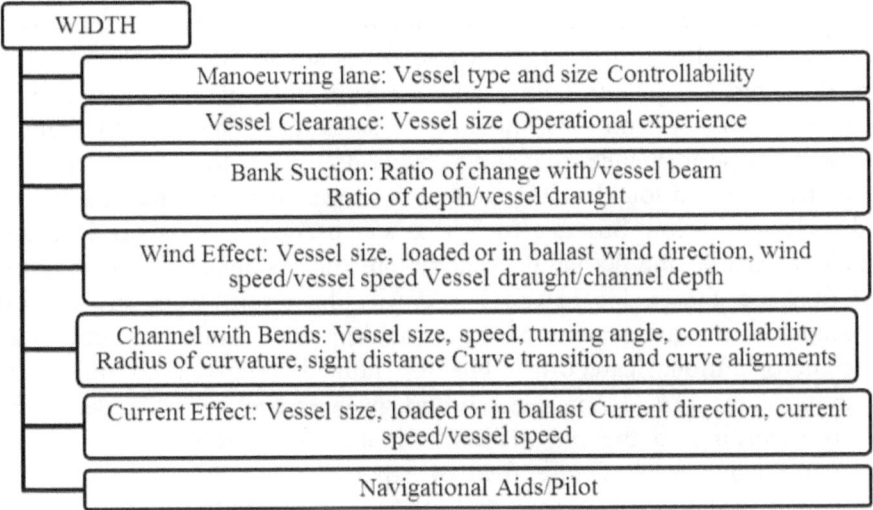

**Figure 3.11**: Waterways width parameters and sub parameters

However, less attention has been given to the channels that will continue to accommodate these ships. Large ships typically maneuver with difficulty in confined areas, and channel width and depth are critical components of deep draft channels. The requirements for access and protection in harbours and ports often lead to maintenance of channels and engineered structures, such as jetties and breakwaters. Table 3.7 shows components of holistic risk assessment that are considered in the risk work.

**Table 3.7:** Component of total assessment

| External impose parameters | Internal control parameters |
|---|---|
| Climate: temperature and humidity | Climate: temperature and humidity |
| Inclination: static and dynamic | Atmosphere: CO2 levels |
| Weather: hall, rain and wind | Flora : mold and fungi |
| Green seas | Shock /vibration |
| Lightning | Communication/ Noise |
| Icing | Flooding |
| Airborne: contaminants and predators | Material |
| Shock: earthquake and explosion | EMC / Lightening |
| Terrorist and piracy | Signature |

# 3.5  SARM Identification of Risk Using HAZID

In qualitative assessment, the design concept needs to address the marine environment in terms of those imposed on the waterways and those that are internally controlled. It is also necessary to address the effects of fire, flooding, equipment failure and the capability of personnel required to operate the system. The qualitative assessment starts with standard and system performance matching. Section 3.5.1.1 to 3.5.1.6 described qualitative method for HAZID. Table 3.8 shows qualitative tools, suitability and their application. Suitable tools that suit particular risk work is chosen, PrHA, HAZOP and What if Analysis are most suitable for new system, and they are chosen for HAZID SARM. The results from the HAZID are recorded in a risk register or matrix stating total number of hazards and different operational categories. The top ranked hazards according to the outcome of the HAZID is selected and given respective risk index or risk matrix according to the on qualitative judgment by the HAZID process and participants from diverse field of expert for the case study (HAZID risk matrix is described in section 3.5.1.6). Qualitative assessment leads to risk ranking where the most serious risk scenario is considered for further analysis. Risk ranking can be index according to level of risk. This is identified and prioritized at the technical workshops for the case study. The following section below describes the respective process. Qualitative analysis is best analyse using combination of suitable tools presented in Table 3.8 as presented in Figure 4.12.

**Table 3.8:** Qualitative risk and reliability tools

| Qualitative Methods | Application |
|---|---|
| Checklist/ Task Analysis/ Historical Incident | Ensure that organizations are complying with standard practice |
| Safety/Review Audit | Identify equipment conditions or operating procedures that could lead to a casualty or result in property damage or environmental impacts. |
| What-If | Identify hazards, hazardous situations, or specific accident events that could lead to undesirable consequences. |
| Hazard and Operability Study (HAZOP) | Identify system deviations and their causes that can lead to undesirable consequences and determine recommended actions to reduce the frequency and/or consequences of the deviations. |
| Preliminary Hazard Analysis (PrHA) | Identify and prioritize hazards leading to undesirable consequences early in the life of a system. |
| FMECA | To assescthe cirticallity of failure |

The analysis is followed by determine recommended actions to reduce the frequency and/or consequences of prioritized hazards.

## HAZOP/FMECA

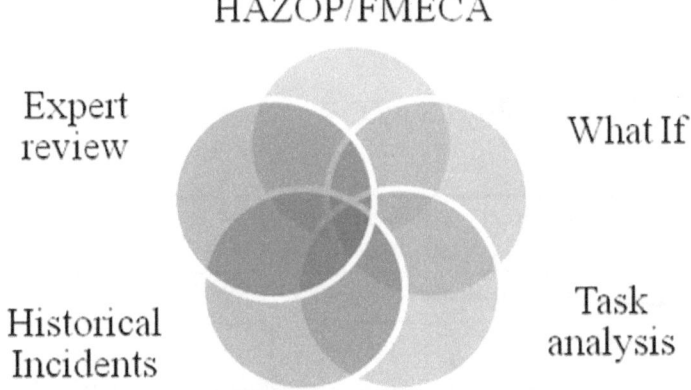

Expert review

What If

Historical Incidents

Task analysis

**Figure 3.12:** Integrated qualitative method

### 3.5.1 Preliminary Hazard Analysis (PrHA)

PrHA represent techniques are part of HAZID to identify hazards that arise from planned/unplanned change. Figure 3.13 shows the process of PrHA.

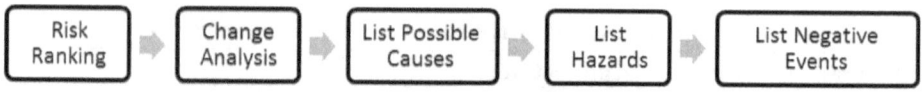

**Figure 3.13:** Flow chart of PrHA

Methodology for PrHA include brainstorming, it can be applied separately or with other tools. In this study PrHA is applied with what if Analysis and HAZOP. PrHA involve review of previous operation/current practices; review operational analysis of planned operation. For each step/phase of the operation, identify differences between the two and determine their impact on risk of the operation. The brain storming involves technique which guides a group in exchanging/generating ideas. It involves:

i.   List negative events, State question and time limit,
ii.  List hazard, Share and record ideas, Discuss ideas to ensure understanding,
iii. Encourage active participation by all, Develop a high-energy, enthusiastic climate, Do not criticize or compliment ideas as they are presented, Encourage creative thinking, Build and expand on the ideas of others, and Try to generate as long a list as possible within the allotted time.
iv.  Change analysis, involve analysis of possible consequence to the system
v.   Risk ranking, based on judgment is made and numbers are assigned to categorise risk.

### 3.5.2 Hazard Operability (HAZOP)

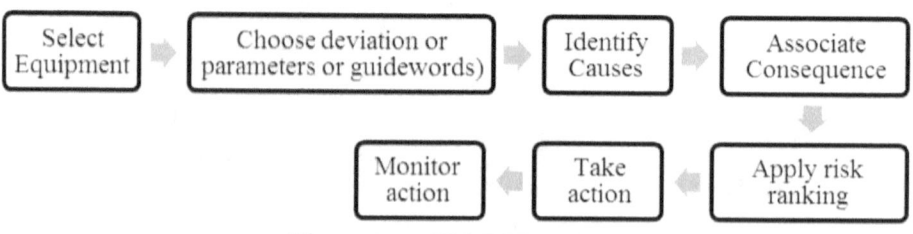

**Figure 3.14:** HAZOP process

HAZOP is done to ensure that the systems are designed for safe operation with respect to personnel, environment and asset. In HAZOP all potential hazard and error, including operational issues related to the design is identified. A HAZOP analysis is detail HAZID, it mostly is divided into sections or nodes that involve systemic thinking and assessment a systematic manner the hazards associated to the operation. The quality of the HAZOP depends on the participants. For the case described in this thesis, representative in the HAZOP meeting are from diverse background in maritime field. Good qualities of HAZOP (HSE, 2001), include politeness and interrupting: i) to the point discussion, ii) Avoid endless discussion iii) be active and positive, iv) be responsible, v) Allow HAZOP leader to lead.

HAZOP involves how to apply the regulation for that process hazard with potential of the major accident. Analysis of for consequence of the process deviation, failure on demand and spurious function of the safety system, alarm function and operator intervention is very important for HAZOP study. Identification of HAZOP is followed with application of combined event tree and fault tree analysis (which are described under uncertainty analysis), probabilistic risk analysis for determination of safety critical elements, training requirement for the operators and integrity and review of maintenance manuals (which is described in quantitative risk analysis). HAZOP process use of guide word / brainstorming, deviation consequence safeguard recommended action. HAZOP record the worksheets efficiently cover all phases. Figure 3.16 shows HAZOP process. HAZOP steps, whose result is tablelated in chapter four is as followed:

i.   Select equipment, in this the most probable system (propulsion failure) that leads to machinery failure is selected.
ii.  Choose deviation or parameters of guidewords for propulsion failure are chosen, this include, no pitch, no rotation.
iii. Identify causes, causes are identifies
iv.  Associate consequence, consequence are deduced
v.   Apply risk, risk index is created
vi.  Take action, required action are deduced
vii. Monitor action are deduced, this indicate whether the chosen solution is working or not.

## 3.5.3 Checklist Qualitative Assessment

The checklists are extracted from International Safety Management (ISM) procedures operation and maintenance (See appendix D). The procedure is matched with current practice and the gap in the system is used to deduce level of risk. Every port and waterways call required the master's and chief officer's intense involvement, like preparations for arrival and departure, pilotage, supervision of cargo operations, official and cargo paperwork. List of activities, tasks or other elements that make up the risk factors are analised under the following categorized together with expected recommended action to be carried out to avid collision according to Collision Regulation (COLREG) guideline:

i.    Arrival Checklist: The test prior to arrival check should be carried out after long ocean passages and before entering restricted coastal areas.

ii.   Departure Checklist: The test prior departure check should be carried out after long ocean passages and before entering restricted coastal areas. Actions to be carried out:

iii.  Incident Command System (ICS) Emergency Checklist Emergency Checklist: The primary duties of the Officer of the Watch (OOW) are watch keeping, navigation and GMDSS radio watch keeping complying all times with the COLREGS and Standard and Watch keeping Certification (STCW95).

iv.   The Sole Lookout Checklist: A sole lookout is allowed only during day light according to the STCW Code. The Master, before allowing has to ensure that it is safe to operate with a sole lookout and therefore carefully has assessed the situation, taking into account at least following relevant actions

v.    The Radar Checklist: Basically radar should be kept running and fully operational at all times.

vi.   The Log Books Checklist: A correct record of the movements and activities of the vessel should be kept in the appropriate log book during the watch. Instructions for the completion of log books should be strictly observed as per respective national regulations and rules.

vii.  The Daily Tests and Checks of Navigational Equipment Checklist: All navigational and radio equipment has to be checked periodically to ensure satisfactory and safe operation.

## 3.5.4 Risk Assessment Ranking and Matrix

Technique designed to assess the risk associated with a hazard, based on severity and probability is shown in Table 3.10. The application is used in any hazard assessment, including hazards identified by multiple sources. Methodology for a given hazard include estimate of hazard severity, estimate of mishap probability and assign risk Assessment Code (RAC), (IMO, 2006). RAC is recommended By IMO. Table 3.9 to 3.12 show risk acceptability criteria recommended for maritime systems. Qualitative risk modeled in this study is measured using IMO risk acceptability criteria given in Table 3.10, Table 3.11. Waterways risk assessed for generic water risk matrix table is also constructed according to according to IMO risk matrix (Table 3.12). Where frequency estimation and consequence estimation of the risk is composed to give idea of the total risk. Qualitative analysis is subjective and the decision depends on the assessor experience.

**Table 3.9:** Frequency probability of occurrence (IMO acceptability criteria)

| Frequency | Probability | Class |
|---|---|---|
| Very unlikely | less than once per 10000 years (P<1/10000) | 1 |
| Remote | once per 100—1000 years (1/1000= P< 1/100) | 3 |
| Occasional | once per 10—100 years (1/100 = P< 1/10) | 5 |
| Probable | once per 1-10 years (1/10=P<1) | 7 |
| Frequently | more than once per year (P =1) | 9 |

**Table 3.10:** Consequence probability of occurrence (IMO acceptability criteria)

| Frequency | Probability | Class |
|---|---|---|
| Minor injury | Less than once per 1000 years (P<1/1000) | A |
| Major Injury | Once per 100—1000 years (1/1000= P< 1/100) | B |
| Death or Total disability | Once per 10—100 years (1/100 = P< 1/10) | C |
| Death total disability for severe person | Once per 1-10 years (1/10=P<1) | D |

**Table 3.11:** Risk matrix

| | Severity | | | | |
|---|---|---|---|---|---|
| | SI | 1 | 2 | 3 | 4 |
| FI | Frequency | Minor | Significant | Severe | Catastrophic |
| A | Frequent | 6 | 6 | 7 | 8 |
| B | Reasonably probable | 4 | 5 | 6 | 7 |
| C | Remote | 3 | 4 | 5 | 6 |
| D | Extremely remote | 2 | 3 | 4 | 5 |

Risk Index = Consequence x Probability, N-Negligible, T-Tolerable, I-Intolerant

Table 3.12 shows another rating based on number that is qualitatively employed for decision.

**Table 3.12:** Traditional Ratings for Occurrence of a Failure

| Rating | Probability of occurrence | Failure probability |
|--------|---------------------------|---------------------|
| 10 | Very high: failure is almost inevitable | > 1 in 2 |
| 9 | | 1 in 2 |
| 8 | High: repeated failures | 1 in 8 |
| 7 | | 1 in 20 |
| 6 | Moderate: occasional failures | 1 in 80 |
| 5 | | 1 in 400 |
| 4 | | 1 in 2000 |
| 3 | Low: relatively few failures | 1 in 15,000 |
| 2 | | 1 in 150,000 |
| 1 | Remote: failure is unlikely | < 1 in 1,500,000 |

## 3.6 Collision Risk Analysis

Upon completion of assessment of entire waterway system analysis described in section 3.3, the collision scenario is specifically assessed and analyised. Figure 3.15 below shows the flow of mathematical model quantitative analysis that is carried out. The analysis is analyzed using excel software and the graphical output is analysed using ALARP principle.

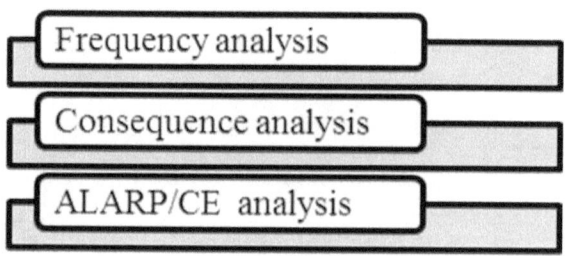

**Figure 3.15**: Risk analysis process

Risk analysis comprises of thorough investigation of accident statistics for specific accident scenario. Risk analysis essentially contains two parts, the frequency analysis and a consequence analysis. A risk model is developed for possible accident collision scenario for IWT. The accident frequency involves estimation of number of occurrence of generic incidents using reasonable accident statistics derived from the selected accident scenarios, the method for this is given in next section and the corresponding result in chapter four. This is compared with similar studies for collision scenario in the waterways of similar profile as well as for other types of ship or by use of statistical or reliability tools, for this the later are used. Whereas, this is not possible, statistical tools are used. Collision model is further broken into different categories as presented in Figure 3.16 shows diagram of components of collision scenarios. SARM is developed for collision prevention model.

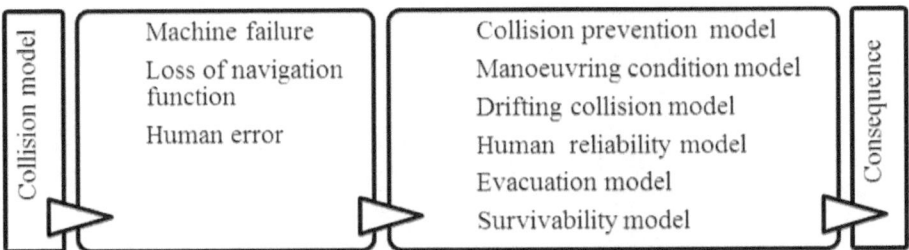

**Figure 3.16**: Components of risk model for collision scenario

The frequency analysis, involve estimation of frequency of generic incidents using reasonable accident statistics derived from the selected accident scenarios.. Surveyed accident statistics and the outcome of the HAZID that lead to generic accident scenarios analysis is followed with recommendation for further risk analysis. This should also be compared with similar studies IWT as well as other ship or use of statistic or reliability technique. Figure 3.17 shows risk analysis index process for collision in waterways. This is determined based on calculation number and numbers obtained from risk curve. Total integrated risk is represented by:

$$R_t = f(R_e, R_s, R_h, R_w) \hspace{2cm} \text{Eq. 3.7}$$

Where: $R_e$ (environment) = f (sensitivity, advert weather . . .), $R_s$ (ship) = f (structural and system reliability, ship layout and cargo arrangement . . .), $R_h$ (crew) = f

(Qualification, fatigue, etc.), $R_w$ =f (other waterways factors).

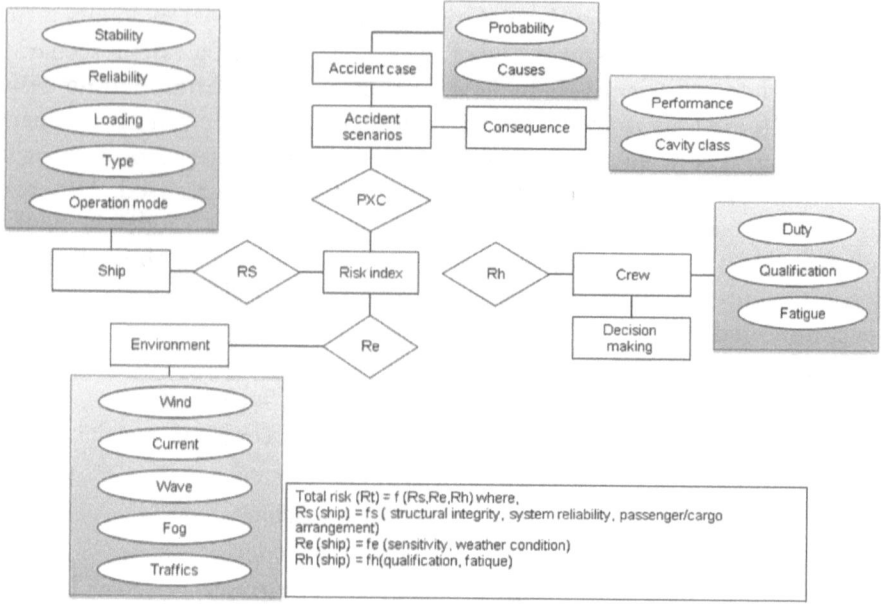

**Figure 3.17:** Total risk analysis index

## 3.7 Analysis of Accident Frequency

Figure 3.18 shows the flowchart of the frequency analysis. The analysis involve mathematical model described below that that involve the following:

i.   Identification of waterway traffic information from waterway system
ii.  Analysis of waterway interaction, using mathematical modeling
iii. Deduction of impact probability from mathematical modeling
iv.  Deduction of accident frequency from the mathematical modeling

**Figure 3.18:** Frequency analysis process

Waterway information is extracted from PIANC data, and the interaction is analysed according to Fuji (1982), the approximate number of ships going around in a waterway analysed by Microsoft excel to deduce corresponding result presented in Chapter 4:

$$N = P(D + B)\rho V_r.$$ Eq. 3.8

where: V is the average speed of the traffic flow, $\rho$ is the average density of the traffic flow, D is the linear cross-section of the obstacle shallower than the draught B is the ship width, D+B is the effective width of the obstacle or shoal. P is the probability of mis-maneuvering or losing control. Absolute system dimension analysis is used for the mathematical model of the system physical representation modeling. The fundamental quantities in absolute system are length (L), time (T), and mass (M). Square brackets "[ ]" are used to show that the dimension of a parameter is being presented.

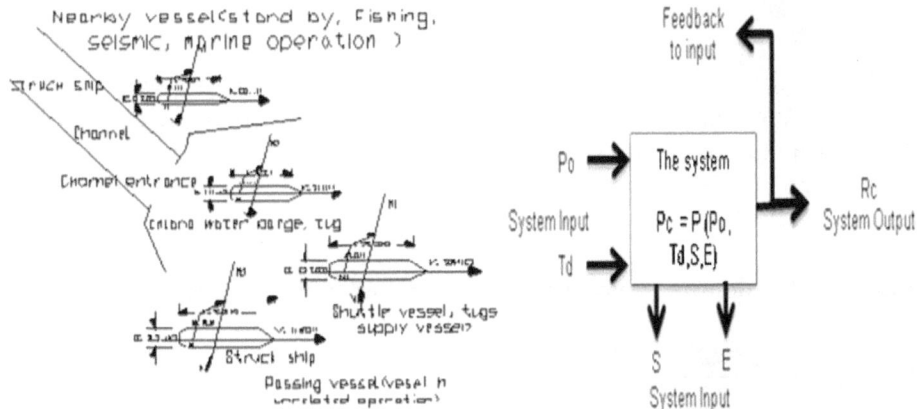

**Figure 3.19a:** Ship movement          **Figure 3.19b:** Waterway physical system

Fuji hypothesis, represent provide the fundamental link between physics and waterway system. It provides the fundamental that lead to physical representation of the channel with required quantity (Figure 3.19).

The dimension of the density of the traffic flow is given by:

$$\rho = \frac{NmXT}{L^2}T_{Er}.$$ Eq. 3.9

[N] = Number of the ships, [D] and [B] = L, [V] = L/T, [P] = Dimensionless, since D is usually much larger than B, B can be ignored.

$$N = PD\rho V_r.$$ Eq.3.10

D is much larger than W, width of the route; the formula then becomes:

$$N = PQ \qquad \text{Eq. 3.11}$$

$$Q = \rho W V_r. \qquad \text{Eq. 3.12}$$

Where: Q is the traffic volume; the dimension of Q (traffic volume) would be gained as:

$$Q = \rho.W = \frac{NmXT}{L^2}.L \qquad \text{Eq.3.13}$$

$$Q = \frac{NmXT}{L}. \qquad \text{Eq. 3.14}$$

The traffic flow density is equal to the traffic volume per unit width of waterway; the traffic volume (Q) should be equal to the product of the traffic flow density ($\rho$) and width of the waterway (W):

$$QQ = \rho\rho \times WW \qquad \text{Eq. 3.15}$$

Taking traffic density is product of flow rate and average speed, we have:

$$Q = \rho.V \Rightarrow \frac{NmXT}{L^2}.\frac{L}{T} \qquad \text{Eq. 3.16}$$

$$\Rightarrow Q = \frac{Nm}{L} \qquad \text{Eq. 3.17}$$

The traffic density is number of the ships per unit area of the waterway. Since traffic for two small boats differs from traffic with two VLCCs1, the dimensions of the vessels is considered in traffic density definition. Therefore a dimensionless parameter as a size factor affected by the traffic density; Therefore the dimension of the traffic density would be:

$$Q = \rho.W = \frac{NmXT}{L^2}.L \qquad \text{Eq. 3.18}$$

$$\rho = \frac{NmXT}{(W.L)Waterway}.\frac{\sum LXB}{(W.L)Waterways}L \qquad \text{Eq. 3.19}$$

$$\rho = \frac{Nm}{L^2}. \qquad \text{Eq. 3.20}$$

Using mass flow rate concept in fluid dynamic, the traffic flow rate is defined as:

$$Q = \rho WV = \frac{Nm}{L^2}.L.\frac{L}{T} = \frac{Nm}{T}$$

Eq. 3.21

Using fluid dynamic concept, $Q$ is the Traffic flow rate representing the arrival frequency of meeting ships where by the desired area is not perpendicular to the path, but is laid on the path (sea surface). Traffic volume or traffic flux is the numbers of vessels navigate through a unit line per unit time.

Traffic volume = traffic density x average speed

Using engineering dimension analysis; the dimension for traffic volume as "the number of the ships per km2" Traffic flow of ships towards a shoal with the effective width of D+B (D and B) average speed of V by Fujii (1982).

$$Q = \frac{Nm}{L^2}$$

Eq.3.22

$$\rho = \frac{NmXT}{(W.L)Waterway} . \frac{\sum LXB}{(W.L)Waterways} L$$

Eq.3.23

$$\rho = \frac{Nm}{L^2} .$$

Eq. 3.24

$$\varphi = \int_{T_1}^{T_2} \rho V(D+B)dt \, \rho. \quad _{=} \phi = (D+B)T.$$

Eq. 3.25

Where: T is the time window that the traffic flow is desired to be calculated in it. If it is so, this calculated traffic flow can be considered as the number of grounding candidates.

$$\phi = (D+B)T. \xrightarrow{B<<D} N_G = \varphi DT$$

Eq.3.26

Traffic density of meeting ship is given by: $N = \varphi DT . XP_C$ Eq. 3.27

The above modeling considers traffic density as constant. If the average traffic density is not used or the time window is not small enough that the traffic density could be considered constant in it, then traffic density ($\rho$) should be defined as a function of time.

$$\varphi = \int_{T_1}^{T_2} \rho(t)V(D+B)dt\,\rho.$$

Eq.3.28

$$\rho_s = \frac{\tau.N}{\tau.W.V} = \rho_s = \frac{N_m}{W.V_1} =$$

Eq. 3.29

Below are representation of various collision situations for head on, overtaking and crossing collision at specific angle. Where: B1 = mean beam of meeting ship (m), V1 = mean speed of meeting ship (knots), B2 = beam of subject ship (m), V2 = speed of subject ship (knots), Nm = arrival frequency of meeting ships (ship/ time), D= relative sailing distance. The generic model based on random collision scenario represents any of the above accident situations including shown in Figure 320 a, b and c. Table 3.13 represents first principle deterministic relation for different collision situation, assuming same speed (Paik, 1996).

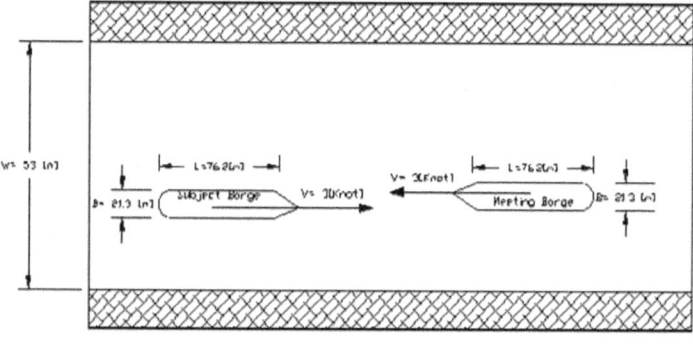

a.

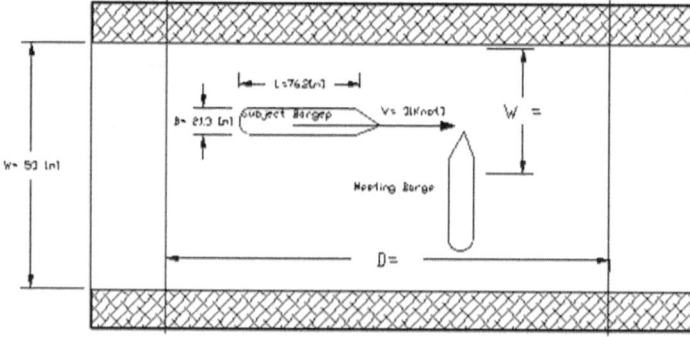

b.

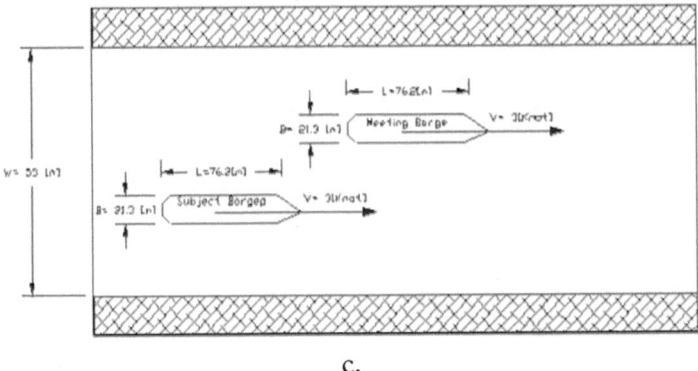

c.

**Figure 3.20 a, b, c:** Collision situations, a. overtaking, b. passing cases, c.crossing

**Table 3.13**: Expression for number of accident impact (Ni)
at different collision situation (Fuji, 1982)

| Expression | Head-on | Overtaking | Crossing | Crossing at specific angle | Circular | Random |
|---|---|---|---|---|---|---|
| Basic | $4xBx$ $Dx\rho_s$ | $\dfrac{(B_1 + B_2)}{W}x$ $\dfrac{(V_1 + V_2)}{V_1 x V_2}x$ | $4x(B + L)$ $xDx\rho_s$ | $\dfrac{dxwx\rho_I}{V_2}x$ $\dfrac{V_2}{\sin\theta} x \dfrac{V_1}{\tan\theta}$ | $\rho_n x Wx \dfrac{8xd_i}{\pi}$ | $N_i = \dfrac{N}{\tau x V}x$ $(\frac{4}{\pi}xL + 2B)$ |
| Standard | | $DxN_m$ | $14xBx$ $Dx\rho_s$ | $xV_i$ | $48xDxBx\dfrac{\rho_n}{\pi}$ | $9.6xDx\rho_n xB$ |
| Relative | 1 | 1 | 3.5 | 3.5 | 3.8 | 2.4 |

Approximations—> L=6B, D=W, $N_i = p_i$

Fairway section is represented with wide characteristics dimension (W), and randomly distributed traffic, where. $P_i$ = probability of collision (C) incident, Se= exposed ship to the encounter ship at period (Te), Vm= speed of entering ship exposed to traffic at period (Tm), $P_n$ = the probability that $(S_n)$ is exposed in the fairway, τ =To= Annual operational time, Nm = Number of ship movement, Ni= The expected number of impacts between incoming traffic or meeting ship $(S_m)$ and existing traffic or subject ship. $(S_n)$, Pnm = conditional probability per unit for collision event that Sn collides with Sm. The probability of collision between entering vessel (Sm) and existing vessel (Sn) can be estimated, where all course is assume to have the same likelihood.

$$T_m = \frac{W}{V_m} =$$

Eq. 3.30

$$p_n = \frac{W}{Ve} \cdot \frac{1}{\tau} =$$

Eq. 3.31

$$N_i = T_M \cdot \Sigma \sum_{E=1}^{E} P_e \cdot P_{em}$$

Eq. 3.32

Considering geometry dimensioning know where is the final meeting conditioned is transformed by aligning Vm direction along x = axis with basis in the origin.

Relative speed

$$V_r = V_e + V_m.$$

Eq. 3.33

$$d_i = d_m + d_e.$$

Eq. 3.34

Figure 3.21 a and 3.21b show random collision situation in a ship movement and in waterway.

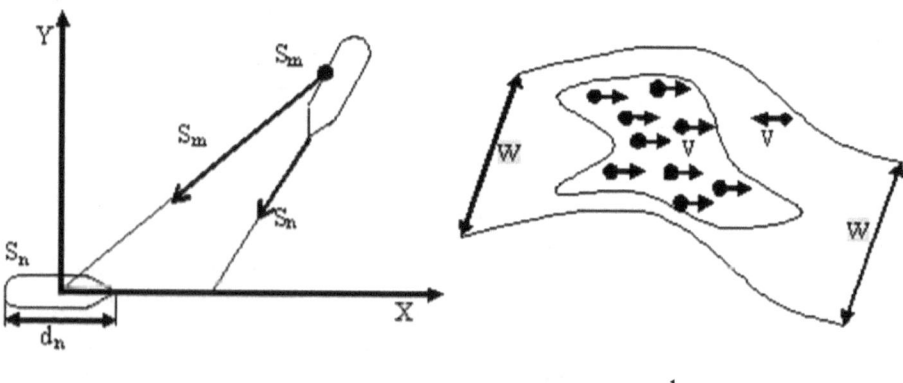

a.                                        b.

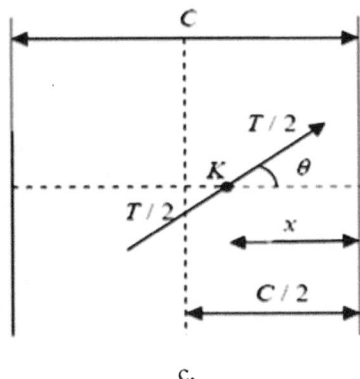

c.

**Figure 3.21: a. , waterway, b. waterway shipc.** Geometry

Pi can be express as the relationship between diameter and characterized width (W)

$$P_I = d \frac{d_i}{w} \; .$$

Eq. 3.35

$$T_E = \frac{W}{V_r} \; .$$

Eq. 3.36

$$\frac{P_I}{T_E} = \frac{d_i V_r}{W^2} \; .$$

Eq. 3.37

$$P_{ME} = \frac{P_I V_r}{T_n} \; . \; = \frac{1}{\pi} \int_0^\pi \frac{d_i V_r}{W^2} . d\theta .$$

Eq. 3.38

The impact diameter di is normal to the relative speed and is given by vector sum. Model is approximated as followed:

Number of impact

$$N_i = \frac{N}{\tau V_r} (\frac{4}{\pi} . L + 2 . B)$$

Eq. 3.39

Assuming the same characteristic for direction and speed, expected number of ship in the fairway at any given time:

$$N_m = (\frac{W}{V}.\frac{1}{\tau})\,N$$

Eq. 3.40

The traffic density

$$\rho_n = (\frac{N_m}{W^2} = \frac{W}{V}.\frac{1}{\tau}.\frac{1}{W^2}) = \frac{N}{V.\tau.W}.$$

Eq. 3.41

By substitution

$$N_i = \rho_n.W(\frac{4}{\pi}.L + 2B)$$

Eq.3.42

Standard Ni for random is given by: $N_i = 9.6.D.\rho_{n}..B$

Where 9.6 is the constant for other viable in the formula

Considering loss of control potential (Pc), accident probability (Pa) is given by:

$Pa = P_c.P_i$ Eq. 3.43

$Na = 2\mu_c.D.N_e P_i$ Eq. 3.44

Where, Na = Pa . . .

$P_c = 2\mu_c.D.N_e$ Eq. 3.45

Causation probability Pc is the probability of failing to avoid the accident while being on a collision. The geometrical collision probabilities got from above analysis is multiplied by a causation probability Pc to get the probability of collision. grounding course. Causation Probability considers combination of important factor affecting factors like environmental and human factors Pc can be estimated from accident data collected at various locations and then transformed to the analyzed area, Pedersen. The advantages of this approach are its simplicity and related robustness.

The causation factor depends on several functions related to traffic perception, communication and avoidance actions. It also depends on external factors such as the vessel types involved in the potential collision situation, weather conditions, studied area and the properties of the ship groups etc. A ship being on a collision/grounding course is called an accident candidate. An accident candidate may result in an accident for example because of a technical fault, or human error. Causation probability quantifies the proportion of cases when an accident candidate, ends up grounding or colliding with another vessel. Primary data for the generic model are extracted from PIANC criteria for waterway classification. PIANC waterway classification is used to select boundary for secondary date (Width, beam, length, draft, and height of bridge) required to select boundary for secondary data PIANC waterway classification is used to select boundary to analyze risk in generic waterways (See Table 3.14)

**Table 3.14:** Waterways classification (ECMT & PIANC, 2008)

| Class | Designation | Max length | Max. Beam | Draught | Min height under bridge | Tonnage |
|-------|-------------|------------|-----------|---------|-------------------------|---------|
|       |             | **(m)**    |           |         |                         |         |
| RA | Open boat | 5.5 | 2 | 0.5 | 200 | |
| RB | Cabin Cruiser | 9.5 | 3 | 1 | 3.25 | |
| RC | Motor yatch | 15 | 4 | 1.5 | 4 | |
| RD | Sailing boat | 15 | 4 | 2.1 | 30 | |
| III | Motor Barge | 38.5 | 5.05 | 1.80-2.20 | 4 | 100-1200 |
| IV | Motor Barge | 85 | 9.5 | 2.5 | 5.25 | 1250-1450 |
| V | Motor Barge | | | | | 1000-3200 |
| VI | Push Tow | | | | | 3200-18000 |
| VII | Push Tow | 285 | 34.2 | 2.50-4.50 | 9.1 | 14500-27000 |

Secondary data used for approximation for probability of losing control are shown in Table 3.15-3.17. Table 4.18 shows ship angle course for collision situation. The probability of failure per nautical and per passage for different collision situation (Pi), failure per nautical mile, per passage ($\mu$) for different waterways and probability of losing control Pc are selected from Table 3.16 and Table 3.17; The probability of loosing navigation control within the fairway failure per passage (Pc) is calculated from this. This study consider random collision situation, therefore the average Pc selected is $2.5 \times 10^{-5}$.

The probability of collision (Pa) is obtained from probability of losing control leading to collision per passage. Collision per annum (Fa) is obtained to estimate accident frequency or collision per number of years. Design of the width of channel depends on vessel Beam (B), Maneuvering lane of channel

(1.8 x B), Ship passing clearance (1 x B) and Bank clearance (0.75 x B for each side (Fuji, 1978) & (Lewison, 1972).

**Table 3.15:** Failure per nautical mile and per passage for collision situation

| Collision Scenario | $\mu_c$ (Failure per nautical miles or hours) | Pc (Failure per passage or encounter) |
|---|---|---|
| Head On | $1.5 \times 10^{-5}$ | $2.7 \times 10^{-5}$ |
| Overtaking | $1.5 \times 10^{-5}$ | $1.4 \times 10^{-5}$ |
| Crossing | $3.0 \times 10^{-5}$ | $1.3 \times 10^{-5}$ |

**Table 3.16:** Failure per nautical mile per passage for different waterways

| Collision Scenario | $\mu_c$ (Failure per nautical miles or hours) | Pc (Failure per passage or encounter) |
|---|---|---|
| Fairway | $2.5 \times 10^{-5}$ | $7.0 \times 10^{-5}$ |
| UK | $1.5 \times 10^{-5}$ | $1.4 \times 10^{-5}$ |
| US | $3.0 \times 10^{-5}$ | $1.3 \times 10^{-5}$ |

**Table 3.17:** Channel bank clearance

| Channel minimum bottom width | 1 Ways traffic Bank Clearance +1 maneuvering lane 2 | 2 Ways traffic maneuvering +1 2 Bank Clearance + 2 lane shipping passing clearance |
|---|---|---|
| | 3.3 B (m) | 6.1 B (m) |

**Table 3.18:** Ship course angle for collision situation

| | |
|---|---|
| Ships Head-on (135°-225°) | $1.5 \times 10^{-5}$ |
| Ships Overtaking (320°-045°) | $1.5 \times 10^{-5}$ |
| Ships crossing (45°-135°), (225°-320°) | $1.5 \times 10^{-5}$ |

Frequency analysis is estimated from traffic density, where expected number of collision per passage (Ni), Necessary period for ship to pass the fairway T is calculated. Primary data used for the model is based on criteria defined by PIANC for waterway classification (See Table 3.15). Table 3.16 and 3.17 represent primary data from previous waterways study. Table 4.18 is the Table of bank clearance and Figure 4.19 is the Table of normally navigation angle course for different collision situation. Random collision can happen in between this entire angle at different collision impact probability (Ni).

This is in according to scenario approach or synthesis approach. Scenario approach is used if Pc is calculated on the basis of available accident data. Pc is estimated from accident data collected at various locations and then transformed to the analyzed area. The advantages of this approach are its simplicity and related robustness. In synthesis approach, specific error situations are supposed to occur in the vessel. They may cause an accident if they take place before or at the same time with a critical situation. Probability of an accident is defined as:

$$P = Ni \times Pc \qquad\qquad\qquad Eq.3.46$$

Where Ni is the geometrical probability or the probability of being on a collision;course. Ni accidents would occur if no aversive maneuver is made. The number of encounter situation (Ne) is given by:

$$Ne = \frac{1}{2} N_m . P_i \qquad\qquad\qquad Eq.\ 3.47$$

Therefore Annual accident frequency (Fa) is represented by:

$$Fa = \frac{1}{2} \mu_c . D . P_i . N_m \qquad\qquad\qquad Eq.\ 3.48$$

$$Fa = P_C . P_i . N_m \qquad\qquad\qquad Eq.\ 3.49$$

Where, W= D. Considering mean sample of the data, traffic density is given by:

$$\rho = \frac{N_m}{v \cdot \tau \cdot W} \qquad\text{Ships/m}^2 \qquad\qquad Eq\ 3.50$$

Expected number of collision: $Ni = 9.6 . B . D . \rho_s = (1/passage)$ Eq 3.51

Necessary period for ship to pass the fairway $T = D/v$ sec Eq 3.52

Therefore, average Pc and $\mu_c$ is considered. (Fuji & Lewison). The probability of loosing navigation control within the fairway

$$Pc = \mu_c . T \qquad\qquad \text{failure/passage} \qquad\qquad Eq\ 3.53$$

Probability of collision Pa= Pi. Pc collision/passage) Eq 3.54

Collision per annum (Fa) = $Pa.N_m$      Eq 3.55

The result of frequency probability Na or F analysis observed can be checked and compared with risk acceptability criteria index shown in frequency acceptability Table 3.19, where by above exponential of five is consider risky, while below exponential 3 is considered negligible. The following are the result of Accident frequency plots from the model after implementation. Table 3.20 shows acceptability criteria for transformable consequence.

**Table 3.19:** Frequency acceptability

| Frequency Classes | Quantification |
|---|---|
| Very likely | Once per 1000 year or more likely |
| Remote | Once per 100-1000 year |
| Occasional | Once per 10-100 year |
| Probable | Once per 1-10 year |

**Table 3.20:** Consequence acceptability

| Fatality | Money | Oil Spill | GHG Emission | Risk classification | Societal Risk acceptability | Risk acceptability per year |
|---|---|---|---|---|---|---|
| Fat | M | OS | $CO_2$ | RC | SRA | RA/Y |
| 0 | 1,000 | <501 | 0~3t/y | (RC-3) | 3—Negligible | 3—Negligible |
| | - | 0.05~1.4 | | (RC-4) | 4—Minor | 4—Negligible |
| | - | 1.4 | | (RC-5) | 5—Significant | 5—Tolerable, |
| 1 | 3,000,000 | 1.4~44 | 3~30 t/y | (RC-6) | 6—Serious | 6—Unwanted |
| 2~10 | 30,000,000 | 44~1400 | 30~300 t/y | (RC-7) | 7—Severe | 7—Unacceptable |
| 10~100 | 300,000,000 | 1400~4500 | 300~30000000 t/y | (RC-8) | 8—Catastrophic | 8—Unacceptable |
| > 100 | 3,000,000,000 | >4500 | - | (RC-9) | 9—Disastrous | 9—Unacceptable |

The outcome frequency (Fo) is then: $F_o = F_e \prod P_b$      Eq. 3.56

Where, $F_a$ is failure frequency, $P_b$ probability of one segment or subsystem Not all possible outcomes can be modeled. Representative scenarios are selected for modeling, and the scenario frequency is taken as:

$$F_S = \sum_{outcomes} F_O$$      Eq. 3.57

Failure per nautical mile and failure per passage can be selected from previous representative work. Necessary period for ship to pass the fairway T=D/v = 3000/3 = 1000 second. The result of accident frequency can be compare with acceptability criteria for maritime industry. If it is too high, the system could be recommended to measure traffic separation scheme (TSS). If the result is high TSS can be model to see possible reduction due to its implementation.

## 3.8 Accident Consequence Analysis (ACA)

Figure 3.22 shows the flowchart of the consequence analysis. It involves:

i. Identification of waterway accident variables from waterways system
ii. Establishment of waterway interaction, using mathematical model
iii. Deduction of energy release probability form mathematical modeling
iv. Deduction of accident consequence from mathematical modeling

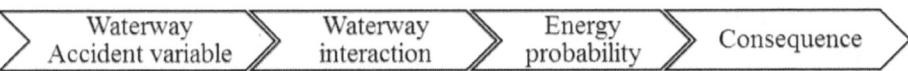

**Figure 3.22**: Consequences analysis process

Waterway accident variables involve consequence resulting from collision remains a big threat to coastal water transportation operation. The nature of the threat can be worrisome. It also helps in the exercise of environmental stewardship for safety and safeguard of environment. Figure 3.23 shows accident scenarios analysis. The following describe the damage modeling of collision accident in waterways Generic risk mitigation option required for operational, societal and technological change decision for sustainable IWT can be deduced from it.

They may lead to loss of life, damage to environment, disruption of operation, injuries, instantaneous and point form release of harmful substance to water, air, soil, water and long time ecological impact. This makes analyzing and quantifying accident frequency and consequence scenarios as function of risk very imperative for reliable design. Vessel width parameter plays a very important role in collision accident scenario and potential consequence damage. Vessel movement for the case under consideration currently has no vessel separation system. However there is traffic movement from both inbound and outbound navigation in the channel. Waterway interaction leading to energy consequence is analysed using mathematical model as followed. The same type of barge size and same speed is considered for the estimation work.

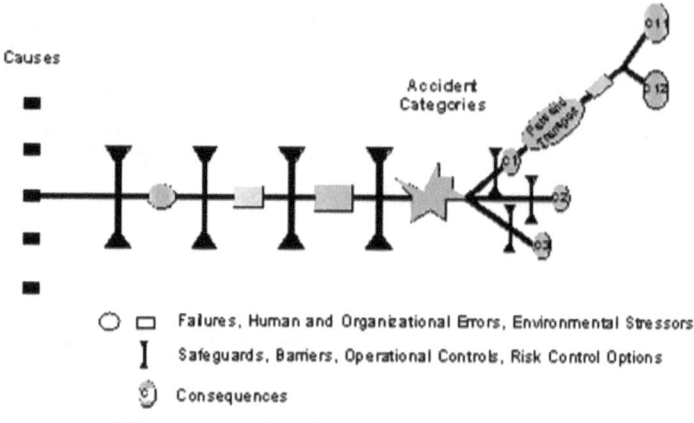

**Figure 3.23:** Accident scenario risk analysis

Initial kinetic energy of ship

$$(Ea) = \frac{1}{2}\frac{M}{1000}V^2.K \qquad\qquad \text{Eq. 3.58}$$

Where: E = impact energy (MJ), M= vessel mass (tonnes), V= vessel speed, K= hydrodynamic added mass constant according to DnV (2000) design code for ship collision.

K= 1.1 for head on collision (powered) impacts, and K= 1.4 for broadside (drifting)

Considering Speed from frequency model $V_f = \dfrac{\rho.W.\tau}{N_m}$    Eq. 3.59

From equation 3.30 (Va) = $\sqrt{\dfrac{2000E_i}{MK}}$          Eq. 3.60

Where $V_f = V_a$

$$\sqrt{\frac{2000E_i}{MK}} = \frac{\rho.W.\tau}{N_m} \qquad\qquad \text{Eq. 3.61}$$

Therefore impact energy that occurs in the accident is given by:

$$Ei = \frac{\rho^2.W^2.\tau^2.M.K}{4 \times 10^6 N_m{}^2} \qquad \text{Eq. 3.62}$$

Absorbed energy can be calculated according to the law of conservation of momentum

$$m_1.v_1.\sin \alpha = (m_1 + (1 + C_h).m_2).v \qquad \text{Eq. 3.63}$$

Ch represents the added mass coefficient of the struck ship and v represent the joint speed perpendicular to struck ship after accident. Assumed loss of kinetic energy in right angle to the struck ship is represented by:

$$\Delta_I = \frac{1}{2} m_1 (v_1.\sin \alpha)^2 - \frac{1}{2}(m_1 + (1 + C_h) m_2) v^2 \qquad \text{Eq. 3.64}$$

This in line with for formula for loss of kinetic energy in a given direction

$$\Delta_I = \frac{1}{2} m_b.v_b - \frac{1}{2}(m_a.v_a) \qquad \text{Eq. 3.65}$$

Inserting joint speed v from equation 3.35 into equation 3.36, we have

$$\Delta E_k = \frac{1}{2} m_1 (v_1.\sin \alpha)^2 - \frac{\frac{1}{2}(m_1{}^2 (v_1.\sin \alpha)^2}{m_1.m_2(1 + c_n)} \qquad \text{Eq. 3.66}$$

$$E_I = \frac{m_1.m_2(1 + C_h)^2}{2(m_1 + m_2(1 + c_h)} (v.\sin \alpha)^2 \qquad \text{Eq. 3.67a}$$

Energy transferred to the struck ship is given by:

$$Ea = \frac{E_I(\frac{1}{(1 + {}^{m_1}\!/_{m_2})})}{} \qquad \text{Eq. 3.67b}$$

$$Ea = \frac{m_1.m_2(1 + C_h)^2}{2(m_1 + m_2(1 + c_h)} (v.\sin \alpha)^2 . \frac{(\frac{1}{(1 + {}^{m_1}\!/_{m_2})})}{} \qquad \text{Eq. 3.68}$$

Absorbed collision impact (Ea) = $47.2V_c$ + 32.8(Minosky, 1959) \qquad Eq. 3.69

Vc= (Ea-32.8)/47.2

Where Vc = collapse material volume (m$^3$). The result from consequence can be compared with risk acceptability for maritime industry. The length of collapsed for damage can be estimated from:

$$L_p = 2.67 \cdot \ln.E_a - 1.97 \cdot \ln 2.8 \left(\frac{M_2}{1000}\right) +1.66 \qquad \text{Eq. 3.70}$$

Alternatively, the consequence analysis, the impact of the scenario on the safety functions is represented by an impairment probability for each safety function (TR, lifeboats; escape routes), which is usually 1 or 0. The impairment frequency for each safety function can be represented by:

$$F_I = \sum_{allscenarios} F_s P_I \qquad \text{Eq. 3.71}$$

Where $P_{I\,is}$ impairment probability of one scenario, the result of frequency and consequence analysis by FN or frequency and damage graph or ALARP analysis as, Risk Cost Control (RCO) assessment using cost of averting fatality index (ICAF) can be carried out. Alternative consequence analysis could focus on measures include risks of safety function impairment, damage risks, annual damage cost, frequency cost (FC) curves, frequency damage size (FD) curves, oil spills, annual spill rate, individual risk, hypothetical individual who is positioned there for 24 hours per day, 365 days per year, and average individual risk, this is usually calculated from historical data.

Individual risk = Number of fatalities/number of people at risk     Eq. 3.72

Individual risk per year and fatal accident rate (FAR) for workers are commonly expressed as a fatal accident rate (FAR). This is the number of fatalities per 108 exposed hours. FARs is typically in the range 1-30, and is more convenient and readily understandable than individual risks per year. FAR can be expressed as:

Onshore FAR = Fatalities at work x 108/person hours at work     Eq. 3.73

To convert individual risk to FAR, the FAR is usually derived from the calculated individual risk of death per year, divided by the number of hours exposed in a year x. the conversion from individual-specific risk to FAR can be represented by:

FAR = Individual—specific risk per year x 108 per 3360 hours per year Eq. 3.74

Alternatively

FAR can be represented by (fatality/year x Hours/year)/hours      Eq. 3.75

Exposure time = Journey/year x Hours

Group risk involves the risk experienced by the whole group of people exposed to the hazard. Societal risks may be expressed in the form of frequency number (FN) or annual fatality rate (AFR). FN curves show the relationship between the cumulative frequency (F) and number of fatalities (N). It constitutes annual fatality rates, in which the frequency and fatality data is combined into a convenient single measure of group risk.

Alternative damage costs estimate could be model using the consequence analysis where damage fraction is estimated for each scenario. This is equal to the cost of the accident as a fraction of the total infrastructure cost. For collision, the consequences of each scenario may be represented by fractions of each module's volume damaged. The total terminal damage fraction (Df) is represented by:

$$D_F = \sum D_m * C_m / C_p$$      Eq. 3.76

Where, Dm = fraction of module damaged, $C_m/C_p$ = cost of module as fraction of terminal cost.

The annual damage fraction is given by:

$$ADF = \sum_{scenarios} F_a * D_F$$      Eq. 3.77

Where, Fs is frequency scenario. The quantity of oil spilled (S), can be estimated as follows: The annual spill rate (ASR) is given by:

$$ASR = \sum_{allscenarios} F_S * S$$      Eq. 3.78

Individual Risks (IR) fatality estimation involves the consequences of each scenario. It is represented by the probability of death for an individual initially a particular location when the event occurs. The overall probability $P_F$ of death is calculated from probability of local fatality in the accident.

$$P_F = P_{fl} + P_{fm}(1 - P_{fl}) + P_{fe}(1 - P_{fm})(1 - P_{fl}) \qquad \text{Eq. 3.79}$$

Where, $P_{fl}$ is probability of fatality during escalation/mustering, $P_{fm}$ is probability of fatality during evacuation $P_{fe}$. The location specific individual risk (LSIR) for a hypothetical individual continuously present in a particular area is:

$$LSIR = \sum_{allscenarios} F_s P_F \qquad \text{Eq. 3.80}$$

Where, $F_s$ is frequency of scenario, $P_F$ is probability of death in the scenario for an individual at the location. The individual specific individual risk (ISRS) for the specific groups of workers from events on the ship is:

$$ISIR = \sum_{allocations} LSIR * P_L \qquad \text{Eq. 3.81}$$

Where, $P_L$ is proportion of time an individual spends in a location. This is the contribution from the modeled events to the individual's total risk. Contributions from external cause like small craft and helicopter flights and occupational accidents can be added in at this stage. The fatal accident rate (FAR) for each group of workers is given by:

$$FAR = \frac{ISIR * 10^8}{H} \qquad \text{Eq. 3.82}$$

Where H is hours of ship movement per year (3360 hours per year is a typical value). For group risks, the total number of fatalities in the scenario is given by:

$$N_F = \sum_{Locations} P_F N_L \qquad \text{Eq. 3.83}$$

Where, $N_L$ average number of people at the location. Thus, for each scenario, a frequency of fatality pair can be obtained. The frequency-fatality and number (FN) curve is the curve of cumulative frequency versus numbers of fatalities. The annual fatality rate is given by:

$$AFR = \sum_{allscenarios} F_S N_F \qquad \text{Eq. 3.84}$$

The result of frequency and consequence analysis is checked with risk acceptability index. This is required to produce influence diagram whose analysis is followed with ALARP principle, cost control option and implied cost of averting fatality index (ICAF). This is followed by risk control option and sustainability balancing of cost benefit towards recommendation for efficient, reliable and effective decision.

## 3.9 Analysis of As Low as Reasonable Possible (ALARP)

Figure 3.23a shows the components of risk assessment and analysis. The analysis leads to risk curve or risk profile. The risk curve is developed from the complete set of risk triplets. Table 3.21shows elements of risk analysi;. The fourth column included shows the cumulative probability, Pi (uppercase P). When the points <Ci, Pi> are plotted, the result is the staircase function. The staircase function can be considered as discrete approximation of a nearly continuous reality. If a smooth curve is drawn through the staircase, that curve can be regarded as representing the actual risk, and it is the risk curve or risk profile that tells much about the reliability of the system. The risk assessment is followed by risk analysis, a systematic process for answering the three questions posed at the beginning the analysis risk and reliability: What can go wrong? How likely is it? What are the impacts? The formal definition of a risk analysis proceeds from these simple questions, where a particular answer is Si, a particular scenario; pi, the likelihood of that scenario; and Ci, the associated consequences. In mathematical parlance, risk triplet (Si, pi, Ci) is shown in Table 3.13 as cumulative risk probability. The analysis that describes and quantifies accident scenario (See Figure 3.23 and 3.24), the risk estimation of the triplets can be transformed into risk curve or risk matrix or risk profile of frequency and consequences plot shown in Table 3.21.

**Table 3.21:** Cumulative risk probability

| Scenario | Probability | Consequence | Cumulative Probability |
|----------|-------------|-------------|------------------------|
| S1 | P1 | C1 | P1=P1+P2 |
| S2 | P2 | C2 | P2=P3+P2 |
| Si | Pi | Ci | Pi=Pi+3+Pi |
| Sn+1 | Pn+1 | Cn+1 | Pn-1=Pn+Pn+1 |
| Sn | Pn | Cn | Pn=Pn |

### 3.9.1 ALARP Principle

Maritime industry risk acceptability criteria established in and to limit the risk is shown in Figure 4.21 and Table 3.24. Table 3.22-3.25 shown ALARP risk acceptability measure; Risk acceptability criteria are established in many industries and regulations to limit the risk in the system Risk acceptability criteria presented in Table 3.22 is used to measure the risk level for the analysis. Consequence thresholds, with priority of value choice are awarded. The highest consequence tripped in order of priority give the overall consequence with descriptor of catastrophic, major and insignificant for 1 People, 2 Infrastructures, and 3 Values (Table 3.21b, Figure 3.24).

Figure 3.17 and 3.18 show domains of experienced fatalities and costs versus annual occurrence probability for different types of engineering structures use as point of reference for measurement of outcome related to cost. Risk is never acceptable, but the activity implying the risk may be acceptable due to benefits to safety, fatality, injury, individual and societal risk, environment, economy. Societal criteria risk that is considered in this research (Table 3.24) is used by increasing number of regulators. Table 3.24 show risk acceptability criteria traditional rating. In the modeling of the potential losses, classification of performance indicators between those phenomena that are ever present during IWTS system lifecycle called systemic phenomenon (probability = 1) and those which occur rarely in the ship lifecycle called random phenomenon (probability < 1) can be made.

| | | | Consequence Criteria | | | | |
|---|---|---|---|---|---|---|---|
| | | | 1—Insignificant | 2—Minor | 3—Moderate | 4—Major | 5—Catastrophic |
| **Likelihood** | A | Consequence certain to occur | Medium (M) | High (H) | High (H) | Very High (VH) | Very High (VH) |
| | B | Consequence likely to occur | Medium (M) | Medium (M) | High (H) | High (H) | Very High (VH) |
| | C | Consequent possibly likely to occur some time | Low (L) | Medium (M) | High (H) | High (H) | High (H) |
| | D | consequence unlikely to occur but could happen | Low (L) | Low (L) | Medium (M) | Medium (M) | High (H) |
| | E | consequence may occur only in exceptional circumstances | Low (L) | Low (L) | Medium (M) | Medium (M) | High (H) |

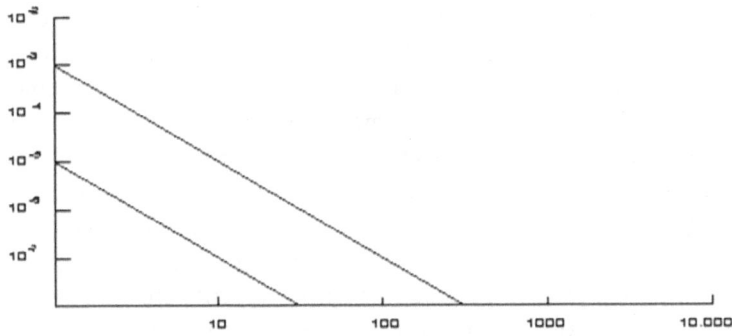

**Figure 3.24:** ALARP Risk Matrix Structure & Risk graph

Consequence Thresholds—priority of value choice is awarded. The highest Consequence tripped in order of priority give the overall consequence. **Catastrophic:** Descriptors of catastrophic consequences for 1. People; 2. Infrastructure; 3. Values. **Major:** Descriptors of major consequences for 1. People; 2. Infrastructure; 3. Values. **Moderate:** Descriptors of moderate consequences for 1. People; 2. Infrastructure; 3. Values. **Minor:** Descriptors of minor consequences for 1. People; 2. Infrastructure; 3. Values. **Insignificant:** Descriptors of insignificant consequences for 1. People; 2. Infrastructure; and 3. Values.

**Table 3.22:** Consequence and fatality Risk

| Risk class | Consequence | Frequent Risk class | | | | |
|---|---|---|---|---|---|---|
| Catastrophic | Ship break apart or loss | 10 | 4 | 5 | 6 | 7 |
| Severe | One or more tanks is penetrated, cargo flow into the sea | 1 | 3 | 4 | 5 | 6 |
| Significant | Cargo thanks are not penetrated, but side or bottom shell plating is penetrated. Fuel oil etc stored in double all hull thanks spill to sea | 0.1 | 2 | 3 | 4 | 5 |
| Minor | No damage to marine environment | 0.01 | 1 | 2 | 3 | 4 |
| | | *Fatality* | *Extremely remote* | *Remote* | *Reasonably probable* | *Frequent* |

A further classification of potential losses can be defined in terms of the consequences impacting the ship or other assets, humans in particular the crew on board), and the environment. Systemic phenomena include: i. degradation of the structure due corrosion, wear, fatigue or increased fuel consumption due to fouling, (assets); ii. Shipboard habitability affected by noise, vibrations or noxious emissions (humans); or iii. Pollutants due to anti-fouling paints or emissions (environment) Random phenomena include: i. Damage or loss to the ship due to collision, grounding, or explosion (assets); ii. Casualty due to personal or ship accident (crew); or iii. Pollution due to oil spill or loss of containment (environment). Table 3.22 shows fatality risk acceptability criteria traditional rating. Table 3.23 give further detail on this classification and the summary of the consequence.

**Table 3.23:** Traditional Ratings for Occurrence, Detection, Severity of Failure (K.-S. Chin *et. al*, Decision Support Systems 2009)

| Rating | Occurrence | | Severity | | Detection | |
|---|---|---|---|---|---|---|
| | Probability of occurrence | Failure probability | Detection | Likelihood of detection by design control | Effect | Severity of effect |
| 10 | Very high: failure is almost inevitable | > 1 in 2 | Absolute uncertainty | Design control cannot detect potential cause /mechanism and subsequent failure mode | Hazardous without warning | Very high severity ranking when a potential failure mode effects safe system operation without warning |
| 9 | | 1 in 2 | Very remote | Very remote chance the design control will detect potential cause/ mechanism and subsequent failure mode | Hazardous with warning | Very high severity ranking when a potential failure mode affects safe system operation with warning |

| 8 | High: repeated failures | 1 in 8 | Remote | Remote chance the design control will detect potential cause/ mechanism and subsequent failure mode | Very high | System inoperable with destructive failure without compromising safety |
|---|---|---|---|---|---|---|
| 7 | | 1 in 20 | Very low | Very low chance the design control will detect potential cause/ mechanism and subsequent failure mode | High | System inoperable with equipment damage |
| 6 | Moderate: occasional failures | 1 in 80 | Low | Low chance the design control will detect potential cause /mechanism and subsequent failure mode | Moderate | System inoperable with minor damage |
| 5 | | 1 in 400 | Moderate | Moderate chance the design control will detect potential cause/ mechanism and subsequent failure mode | Low | System inoperable without damage |

| 4 | | 1 in 2000 | Moderately high | Moderately high chance the design control will detect potential cause /mechanism and subsequent failure mode | Very low | System operable with significant degradation of performance |
|---|---|---|---|---|---|---|
| 3 | Low: relatively few failures | 1 in 15,000 | High | High chance the design control will detect potential cause/ mechanism and subsequent failure mode | Minor | System operable with some degradation of performance |
| 2 | | 1 in 150,000 | Very high | Very high chance the design control will detect potential cause/ mechanism and subsequent failure mode | Very minor | System operable with minimal interference |
| 1 | Remote: failure is unlikely | < 1 in 1,500,000 | Almost certain | Design control will detect potential cause/ mechanism and subsequent failure mode | None | No effect |

The systemic events occur with probability of one, the consequences that follows are still uncertain. The systemic events may as well as the random events be treated in a risk based design as well as summarizes the elements which are also dynamic economics, society, and environment is imperial to be taken into consideration. The risks associated with these elements are both systemic phenomenon (p = 1), or random phenomenon (p < 1). Design has conventionally been concerned with systemic economic risks through first and operating design costs, while owner's hedge their accidental risks via insurance. After decision based on ALARP consideration other channel complexity and uncertainty, reliability and sustainability analysis, consider for adjustment on waterways requirements. Regulation is concerned with the risks to life, the property of third parties and environmental protection. Only in recent years has the systemic risk to the environment become a significant concern, with ever increasing importance (Refer to Table 3.24).

**Table 3.24:** Holistic Risk matrix acceptability

| Entity | P=1 (systemic) | P<1 (random) |
|---|---|---|
| Economic | First & operating costs (Design concern) | Loss, damage of vessel (Owner's concern hedged via insurance) |
| Society | Quality of Life / Corporate Social Responsibility (Regulatory concern) | Loss of Life (Regulatory concern) |
| Environmental | Environmental Impact "Sustainability" (New concern) | Pollution (Regulatory concern) |

## 3.9.2 Risk Acceptability Criteria

Risk acceptability criterion is dynamic because of differences in environment and diversity in industries and choice of regulations to limit the risk. Risk is never acceptable, but the activity implying the risk may be acceptable due to benefits of safety to reduce fatality, injury, individual and societal risk, environment, economy improvement.

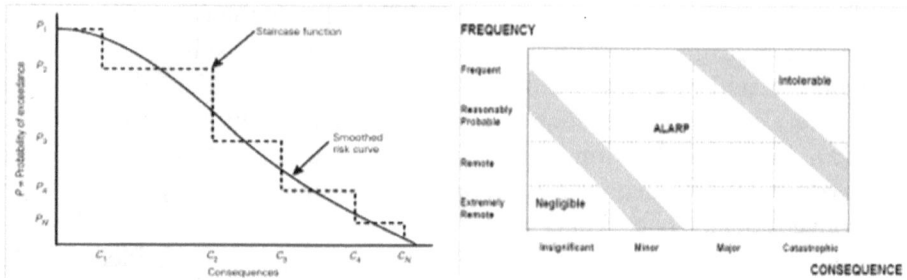

**Figure 3.25a:** ALARP Influence diagram    **Figure 3.25b:** Stair case risk curve

Societal risk acceptability criteria are used by increasing number of regulators. Accident frequency modeling can be on the cargo leakage, the survivability model and the evacuation model related to collision, and grounding. The consequence could further be broken down into effect for ship, human safety, oil spill, damage, ecology, emission uinfrastructure and other environmetal impacts. Further effort on measures to reduce accident can be attempt through sub system level analysis (Figure 3.25a and b). Preceding risk control option and sustainability balance work towards recommendation for efficient, reliable and cost effective decision. The frequency (F) of accidents involving consequence (N) or more fatalities may be established in similar ways as individual or societal risk criteria. For risks in the unacceptable and intolerable risk region, the risks should be reduced at any cost. The risks should be reduced to a level ALARP level. See Table 3.25.

**Table 3.25:** Risk acceptability matrix

| Frequency of Occurrence (or Likelihood) | Consequences (Severity of Accident) | | | | |
|---|---|---|---|---|---|
| | Incidental (1) | Minor (2) | Serious (3) | Major (4) | Catastrophic (5) |
| Frequent (5) | M | H | VH | VH | VH |
| Occasional (4) | M | M | H | Risk without measure | VH |
| Seldom (3) | L | M | H | H | VH |
| Remote (2) | L | L | Risk after measure | H | H |
| Unlikely (1) | L | L | M | M | H |

FN curves are influence diagram of frequency-fatality plots, showing the cumulative frequencies (F) of events involving N or more fatalities.

They are derived by sorting the frequency fatality (FN) pairs from each outcome of each accidental event, and summing them to form cumulative frequency-fatality (FN) coordinates for the plot. FN curves are graphical measures of group risk that show the relationship between frequency and size of the accident. Plot within the ALARP range, CEA or Cost Benefit Analysis may be used to select reasonably practicable risk reduction measures.

### 3.9.3 Alternative Consequence Quantification (Cost of Damage or Loss of Life Quantification)

Accident consequence is dynamic; deduce consequence energy of collision is translated to oil spill, carbon dioxide released and number of fatality or injury( Figure 3.26). Ship collision is rare and independent random event in time. The event can be considered as poison events where time to first occurrence is exponentially distributed. Damage could be to the structure, or environment or person. Table 3.27 provides alternative quantification transformation.

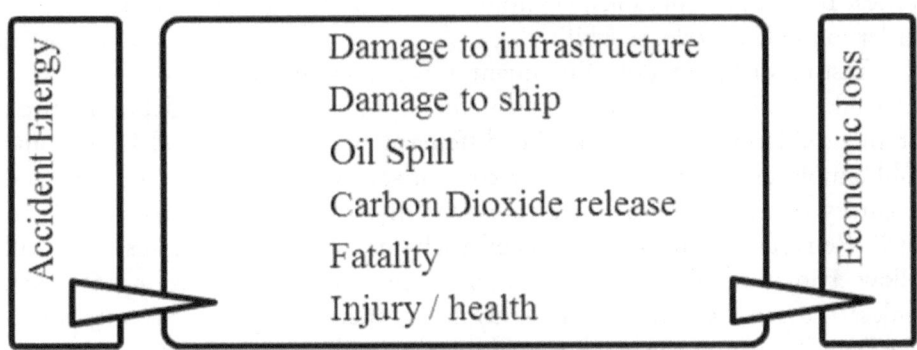

**Figure 3.26:** Other consequence

## 3.10 Analysis of Reliability

Figure 3.27 shows Reliability test process. The process involves:

i.   Compilation and analysis of result data
ii.  Testing of result accuracy on reliability software. In this case minitab.
iii. Observation of the trend for accuracy of statistical inference

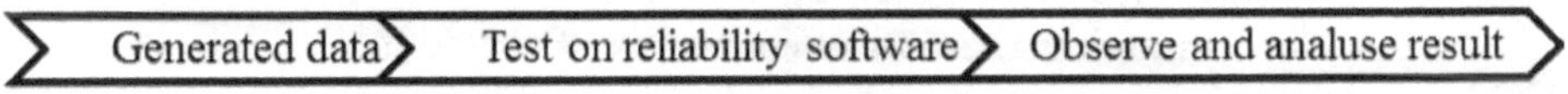

Generated data > Test on reliability software > Observe and analuse result >

**Figure 3.27:** Reliability test process

Validation work required compilation of system. The current state of practice often attempts to validate system models directly from test data taken on the entire system. This approach can be problematic if there are a large number of components containing complex connections or interfaces, energy dissipation mechanisms, or highly nonlinear behavior. There is hierarchy of system on which model can be built and validation carried out on. In this Hierarchy, the top tier is the complete system. Fundamentally model verification and validation work require modeling, simulation and assessment of the system. However, unit problem, component problem, subsystem, or the complete system can also be assed. Waterway system development includes the channel requirement, vessel requirement, environmental requirement, infrastructure requirement and all supporting structure. Waterway model system is further based on three more tiers which are subsystems, components, and unit problems. These tiers illustrate the breaking of a complex system into a series of fundamental physical problems. The number of system breakdown could be more depending on system and associated use. Uncertainty and error in experimental data include variability in test fixtures, installations, environmental conditions, and measurements in order to predict model reliability.

Testing of the model using quantitative comparison of models outcomes may take the form of a statistical statement of the selected validation metrics. If the validation metric was the difference between the models outcome and simply error, the quantitative comparison would quantify the expected accuracy of the model as 95% confident that the error is between 5% and 10%. Reliability on measurements of small number of the parts of system will allow estimation of a mean and standard deviation for distribution. However, unless the sample size is sufficiently large, there will be uncertainty about the "true" values of the statistics, and, indeed, uncertainty regarding the true shape of the distribution. Obtaining more information will allow reduction of this uncertainty and a better estimate of the true distribution.

## 3.11 Cost Effective Analysis (CEA)

This CBA is then followed by assessment of the control options as a function of their effectiveness against risk reduction. In estimating RCO, the following are taken into consideration: DALY (Disability Adjusted Life Years) or QALY (Quality Adjusted Life Years), LQI (Life Quality Index), GCAF (Gross Cost of Averting a Fatality), NCAF (Net Cost of Averting a Fatality). The common criteria used for estimating the cost effectiveness of risk reduction measures are NCAF and GCAF which can be calculated with the following equation:

$$\text{Gross CAF} = \frac{\Delta C}{\Delta R} \qquad\qquad \text{Eq. 3.85}$$

$$\text{GCAF} = \frac{\text{Cost - Benefit}}{\text{Reduction}} \qquad\qquad \text{Eq. 3.86}$$

$$\text{NET CAF= NCAF} = \frac{\Delta C - \Delta B}{\Delta C} \qquad\qquad \text{Eq. 3.87}$$

$$\text{NCAF} = \text{GCAF} - \text{Change in Benefit} \qquad\qquad \text{Eq. 3.88}$$

$$\text{ICAF} = \frac{\text{NCAF}}{R} \qquad\qquad \text{Eq. 3.89}$$

Where: $\Delta R$ is Reduction in annual fatality rate, $\Delta R$ is Economic benefit resulting from implementing the risk control option, $\Delta R$ is Risk reduction in term of averted number of fatality implied by the risk control option. NCAF and GCAF depend on the following criteria:

i.   Observation of the willingness to pay to avert a fatality;
ii.  Observation of past decisions and the costs involved with them; and
iii. Consideration of societal indicators such as the Life Quality Index

In RCO, It is important to address the following:

i.   Primary cause or accident scenario, number of accident,
ii.  Number of losses, number of life loss per accident,
iii. Cost of fatality per accident, average total cost per accident.

$$\text{Cost per unit risk reduction (CURR)} = \frac{\text{Cost - Benefit}}{\text{Reduction}} = \frac{\text{NPV}}{\text{Reduction}} \quad \text{Eq. 3.90}$$

Where 50 minor injuries = 10 serious injuries = 1 life = property or damage = loss or degradation of environment. The NPV can be calculated from:

$$\text{NPV} = \Sigma_{t=1}^{n}[(C_{t+}B_{t})\,(1+^{(1 + r)^{-t}}] \qquad\qquad \text{Eq. 3.91}$$

Or

$$R\left[\frac{(1+i)^{n} -1}{i.(1+i)^{n}}\right] \qquad\qquad \text{Eq. 3.92}$$

Where: t = Time horizon for assessment, starting in year 1, Number of year in vessel life time, B = the sum of benefit in period, r = the discount rate per period, Ct—sum of cost in period. The estimated risk is represented by:

$R_0$= Accident frequency Fa (Number of ships per year) x
Consequence Ca x Co (Cost of damage per accident)     Eq. 3.93

Risk after implementation of safety measure

$R_1$= Accident frequency Fa (Number of ships per year) x
Consequence Ca x Co (Cost of damage per accident)     Eq. 3.94

Benefit of reduced risk (R) = $R_0$ - $R_1$     Eq. 3.95

NPV of the benefit for estimated risk and implemented safety measure is calculated and ratio of cost of C to benefit B is compared and expected to be < 1.

### 3.11.1 Cost of Averting Fatality (CAF)

ICAF represent estimation of benefit of avoiding damage or fatality. It plays important role in cost benefit analysis of risk. ICAF can be estimated using the following means (DnV, 2005):

Ronold Life quality index (L) = $\gamma^w \cdot \varepsilon^{1-w}$     Eq. 3.96

Where: L = life quality index, $\gamma$ = Gross domestic product per person per year, $\varepsilon$ = Life expectancy (year), w = Proportion of life spent in economic activities in developing countries is approximately 1/8.

$$\frac{\Delta\varepsilon}{\varepsilon} > \frac{\Delta\gamma}{\gamma} \cdot \frac{w}{1-w}$$     Eq. 3.97

Where $\Delta\varepsilon = \frac{\varepsilon}{2}$ = ½ of life expectancy, largest change in GDP, $|\Delta\gamma|_{max}$ = - y. (1-w) /2 w.

Optimal acceptable ICAF => $|\Delta\gamma|_{max} \cdot \frac{Y-\varepsilon}{4} \cdot \frac{1-w}{w}$     $\Delta\varepsilon = \frac{Y-\varepsilon}{4} \cdot \frac{1-w}{w}$     Eq. 3.98

Where, Social cost = NC$(1 + i)^t$, t<6000, Social cost = NC $(1 + i)^{6000}$, t>6000, N = number of injuries or fatalities, C = Cost of damage per day depends on types and countries, I = daily rate of interest, T = Duration of damage or sick leave in day, 6000 days is equivalent with a fatality, DNV = US$ 3 million = cost effective ICAF rate = 2GBP million = developed country.

$f(t) = \gamma e^{-\gamma t}$ Eq. 3.99

Where: $\gamma$ = Annual rate of exceeding of consequence energy capacity, t = the time to the future loss

$C_o = C_f . \int_0^T e^{-it} . \gamma^{-dt} . dt$ Eq. 3.100

or

$C_o = \frac{C_f . \gamma}{i+\gamma}$ Eq. 3.101

Total cost $C_t$ = present value of future cost $(C_o)$ + Cost of protective measure $(C_c)$

$C_t = C_o + C_c$ Eq. 3.102

In prioritizing alternative under (RCBA), it is important to address the following: Available concept, consequence energy capacity (MJ), Return for exeedance T (Year) as well as:

Annual rate of exeedance $(\gamma = \frac{1}{\tau}(-))$ Eq. 3.103

And $C_o$, $C_c$ and $C_t$ (US$) Eq. 3.104

The cost effective risk reduction measures should be sought in all areas. It is represented by followed:

Acceptable quotient = Benefit/(Risk/Cost) Eq. 3.105

The dominant yardstick in risk studies is the $3m criterion as described in MSC 78/19/2. According to this, in order to recommend an RCO for implementation (covering risk of fatality, injuries and ill health) this must give a CAF value –both NCAF and CGAF—of less than $3 million.

The criterion of $3m will result in the recommendation of the RCO to be introduced if this is not the case, the RCO is rejected. .This means that for a specific RCO to be adopted, the three variables, namely $\Delta C$, $\Delta B$, and $\Delta R$, have to satisfy the following inequality:

$\Delta C < \$3m\ \Delta R + \Delta B$

For the GCAF criterion, the equivalent inequality is: $\Delta C < \$3m\ \Delta R$

If $\Delta B > 0$ (a reasonable assumption if the RCO in question will result to some positive economic benefit), then if the RCO satisfies the GCAF criterion ($\Delta C < \$3m\ \Delta R$), it will always satisfy the NCAF criterion as well ($\Delta C < \$3m.\Delta R + \Delta B$). 60 000 as the cost of averting one tonne of spilled oil (CATS).

## 3.11.2 Cost of Optimal Energy Capacity

Benefit of River alignment, implementation of Traffic Separation Scheme (TSS), dual propulsion shaft, VTS, AIS or bridge clearance can be quantified into cost of safety and protection of environment (IMO, MSC 2001). Energy capacity that can be absorbed catastrophic energy without collapse is $10^6$ MJ @ Catastrophic collapse. Where, economic loss of bridge structure + loss for use of infrastructure = US$ 2 X $10^8$, I= rate of interest = 3% per year.

Present value of loss=> $C_o = C_f \cdot e^{-it}$ 　　　　　　　　　Eq. 3.106

Where: $C_o$ -Present value of future loss, $C_f$ - loss at deficient time in future or future loss in present monetary unit (notificated), i =real interest rate of future loss, t= the time to the future lost. At random variable and occurrence time, loss is stochastic:

$$C_o = C_f \cdot \int_0^T e^{-it} f(t).dt$$ 　　　　　　　Eq. 3.107

E (t) = the expected value function, f (t) = probability density function for time t), the time to the occurrence of catastrophic collapse. Where: $C_f = 2.10^8$ and I at 3% = 0.03

A relationship between safety, environmental risk components and ageing or failure factor of IWT should be expected achievement of the analysis. Based on this relationship a generic model based on system engineering, scientific and stochastic analysis useful for decision support on damage, maintenance and management of reliable, safety and environmental friendly IWT can be developed.

### 3.11.3 Cost of Safety and Consequence

Sustainability is defined as development work that meets needs of
the present generation without compromising the ability of the future
generations to meet their own needs. It requires balancing work between
technical, developments, economic, community participation, information
sharing, environment and safety. Suitability principle calls on all fields of
human activities to review and adjust the way things are done. At its 21st
session in February 2001, the UNEP governing council adopted a decision
to investigate the feasibility of a "Global Assessment of the State of the
Marine Environment" UNEP GC Decision 21/13. Figure 3.28 shows typical
sustainability analysis graph that can be for decision support for the system.
It was recognized that a global marine assessment (GMA) was needed and
feasible. The scope of an assessment process was outlined. It was agreed
that GMA activities should include cost, technical, environmental, socio
economic considerations, together with the relevant work, approaches and
experience of national, regional and global so that the project in question
can touch more of living thing and environmental consideration and impacts
(UNEP, 2008). The cost of each safety and reliability of each waterway
variables is analyzed.

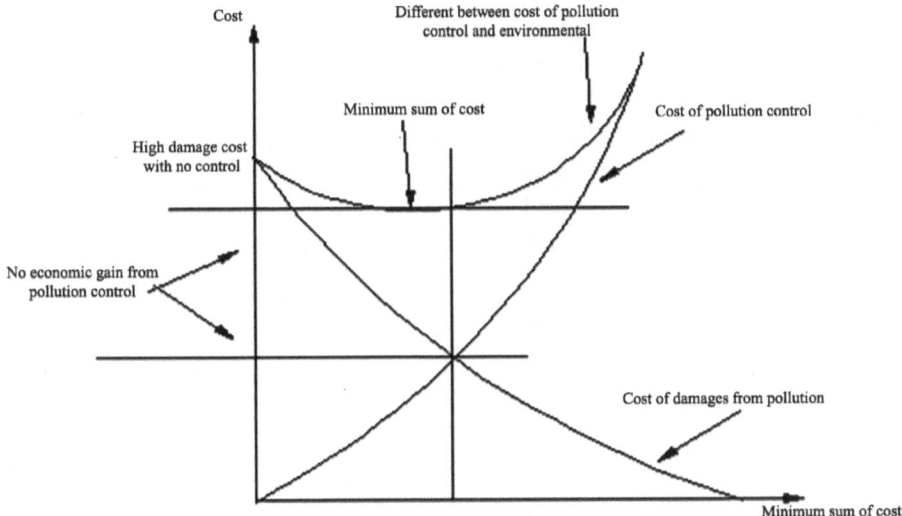

**Figure 3.28:** Sustainability balance

The case presented in this thesis is examples of the daily practice of environmental management. Being anything but a straightforward application towards reliable management work derived from sound decision making which involve judgment and criteria. In order to interpret, assess pros and cons, measure the different interests possibly confronted, and finally choose the best decision according to social interests and the common good. Table 3.26 shows each type of cost for each type of impact from ship accidents. Table 3.27 shows consequence transformation Table between consequences. Figure 3.29b show experienced fatalities and costs versus annual occurrence probability for different types of engineering structures. From Table 3.27 and Figure 3.29a consequence dynamic analysis and translation is done for oil spill and COx from cost and energy analysis, cost for impact from ship accidents. The direct and indirect costs to the ship owner and other affected parties e.g. ports, governments etc and intangible costs to people and the environment are passed back to society.

**Table 3.26:** Categorization of direct and indirect costs of an accident

|  | Direct costs | Indirect costs | Intangible |
|---|---|---|---|
| **Property** | Repair<br>Cargo loss | Off hire<br>Differed production (cargo owners)<br>Lost share value<br>Lost market share | Lost reputation<br>New regulations |
| **People** | Medical treatment | Differed production (sick leave) | Grief and suffering |
| **Environment** | Clean up/ restoring | Lost business (e.g. tourism or fishing) | Ongoing damage to the environment/ecosystem (loss of biodiversity) |

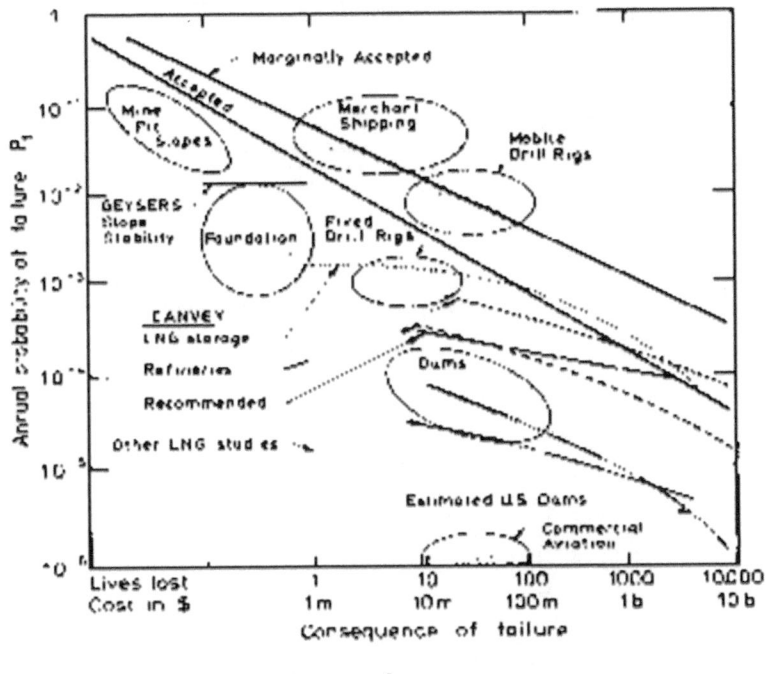

a.

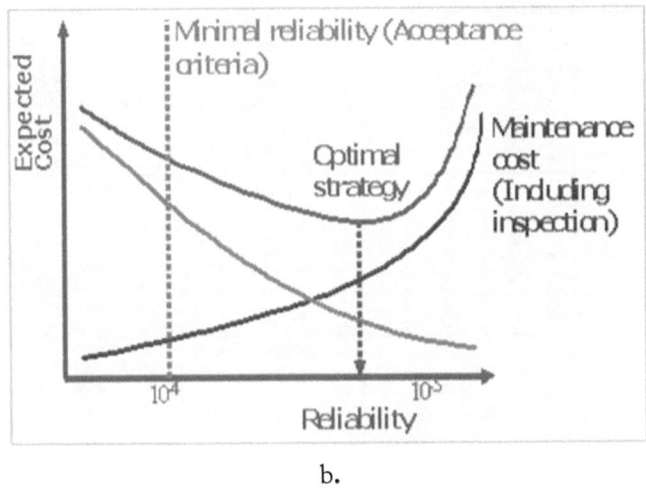

b.

**Figure 3.29:** Measure of fatalities and costs versus annual occurrence probability for different types of engineering structures.

**Table 3.27:** Consequence risk society transformation

| Society | | Abbreviation | Negligible 3 | Minor 4 | Significant 5 | Serious 6 | Severe 7 | Catastrophic 8 | Disastrous 9 |
|---|---|---|---|---|---|---|---|---|---|
| Consequence class | | | | | | | | | |
| Human | | HL | - | - | 5 | 1 fatality | 2-10 fatalities | 10-100 fatalities | > 100 fatalities |
| | | HI | Bruises and minor damages that do not require hospital treatment | 1 injury requiring hospital treatment | Several incidents requiring hospital treatment | Several incidents requiring hospital treatment. Disabilities | - | - | - |
| | | HW | Medical treatment first aid cases | Restricted work accidents— An accident where an individual is unable to perform normally assigned work functions for a period | An accident where an individual is unable to carry out any of his duties or return to work on a scheduled work shift on the day following the injury. 1 disabled | - | - | - | - |

| Environment OS (Ton) | - | <501 | 0.05 ~1.4 | 1.4~44 | 44~1400 | 1400~45000 | >45000 |
|---|---|---|---|---|---|---|---|
| EM | - | Environment Damage | | | | | |
| | | Negligible | Minor | Serious | | Critical | Catastrophic |
| | | Restored within : None | days | weeks | month | 1~2 years | Several years |
| $CO_2$ | 0-3 t/y | 3-30 t/y | 30-300 t/y | 300-3 mil t/y | | | |
| Monetary value (Euro) | 1 | 10 | 100 | 1,000 | 10,000 | 100,000 | 1,000,000 |
| Acceptability per year | Negligible | | Tolerable | Unwanted | Unaccepted | | |

## 3.11.4 Analysis of Risk Control Option (RCO)

Figure 3.30 shows RCO process. It involve

   i.  Reinvestigation of regulation and system functionality
   ii.  Deduction of Risk Control Option from risk analysis result

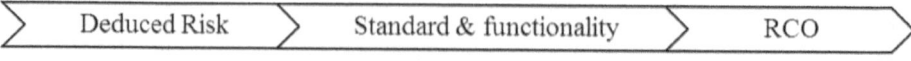

**Figure 3.30:** RCO process

System functionality and standard previously described is qualitatively revised with deduced risk (See appendix D). Risk control measures are used to group risk into a limited number of well practical regulatory and capability options. Risk Control Option (RCO) aimed to achieve preventive: reduce probability of occurrence and mitigation: reduce severity of consequence. RCO could follow general approach: controlling the likelihood of initiation of accidents be effective in preventing several different accident sequences; and distributed approach: control of escalation of accidents and the possibility of influencing the later stages of escalation of other unrelated, accidents.

The economic benefit and risk reduction ascribed to each risk control options is be based on the event trees developed during the risk analysis and on considerations on which accident scenarios would be affected. Estimates on expected downtime and repair costs in case of accidents should be based on statistics from shipyards or responsible government institution for repair or construction. In RCO regulatory framework is check with system functionality, design facto to reduce accidents deduced.

Cost effectiveness analysis is use to deduce mitigation, options selection and proposed need for technology, reliability, new regulations and sustainability required to be modeled for effective mitigation options. RCBA involves quantification of cost effectiveness that provides basis for decision making about identified RCO.

This includes the net or gross and discounting values for cost of equipment, redesign and construction, documentation, training, inspection maintenance drills, auditing, regulation, reduced commercial used and operational limitation (speed, loads). Benefit could include reduced probability of fatality, injuries, serenity, negative effects on health, severity of pollution and economic losses. Identified types of cost and benefits for each risk control option according to RCBA for the entities which are influenced by each option can be deduced. And also identification of the cost effectiveness expressed in terms of cost per unit risk reduction.

## 3.12 Validation of Model by Case Study

Figure 3.31 shows the case study process. The process involves:

   i.   Organization of case study data and identification of parameters
  ii.   Insertion into the model developed
 iii.   Deduction of result for the case study from ii, is given in chapter Four
 iv.   Analysis of the result of the case study in iii is discussed in chapter iv

**Figure 3.31:** Case study validation

Case study data are described in the following session. The case study data are arranged similar to data used for the generic model. For example for the case of Langat rives the River profile, bathymertry and watershed data are the first main background data from where the remaining data for analysis are extracted. The Langat River originates from the main range and drained by numerous tributaries found in the Hulu Langat forest reserve. The river drains the northern and western part of Hulu Langat district, all the way down to Dengkil, Klang Valley area where it meets major tributary the Semenyih River as in Figure 3.40. The major tributaries of Sungai Langat are the Sungai Semenyih and Sungai Labu. The Klang – Langat River watershed represents the most urbanized region in Malaysia. The climate of the study area is equatorial monsoon. Temperature is uniform but high throughout the year. Dry and wet seasons are not particularly well marked, because heavy rainfall may be experienced anywhere at any time of the year. The reason for choosing examples of maintenance of Klang River for the case study is because of recent environmental pressure emanating from global warming, environmental calamity, climate change, and population rise leading to congestion in the city and point form pollutant release. Figure 3.32 shows Google map of Langat River.

**Figure 3.32**: Langat River (Google map)

The bathymetry offshore the estuary is determined by constant activity of dredging due to the presence of the harbour. Bathymetry map (new edition 1998) shows that the depth at the mouth of the estuary is shallow ranging from 1 to 5 meters deep. While at mid channel depth increases drastically to 6 to15 metres. The two types of depth have been maintained by the port authority to ensure sea traffic viability plying the channel towards the Klang South Port.

### 3.12.1 Background of the Langat River

The Langat River originates from the main range and drained by numerous tributaries found in the Hulu Langat forest reserve. The river drains the northern and western part of Hulu Langat district, all the way down to Dengkil, Klang Valley area where it meets major tributary the Semenyih River. The catchment area of Langat River at this confluence is 1240 km and flows out of the Klang valley area to the Straits of Melaka. The Klang River drains an area of about 1200 sq. km extending from the headwaters in the steep mountain forests of the main range of Peninsular Malaysia, to the river mouth for a total length of 120 km. The main tributaries of Klang River are Batu, Gombak, Ampang, Kerayong and Damansara rives. The river system drains all the major urban centers in

Klang Valley, viz. Kuala Lumpur, Petaling Jaya, Subang Jaya, Shah Alam, Klang and Port Klang. The catchments of Klang, Batu and Gombak rivers are relatively small (DOE, 2008). Table 3.28 and Figure 3.33 show Langat River profile.

**Table 3.28**: Langat River base line and river profile data

|  | **Longitude** | **Latitude** |
|---|---|---|
| Basin location | 1010 17' and 1010 53'W | 20 40' and 30 16'N |
| Length | 120—137 km long—originates from the Main Range and drains westward to the Straits of Melaka | |
|  | North Estuary | South Estuary |
|  | 44.2 km | 9.9 km |
| Coverage | Hulu Langat | Clang river |
|  | (81 779 ha), Seeping (61 621 ha) and Kuala Langat (78 598 ha) | districts of Clang (61 169 ha), Pedaling (47 626 ha), Gombak (60 372 ha) and Kuala Lumpur (24 612 ha) |
| Total area | 415 779 ha square meter | |

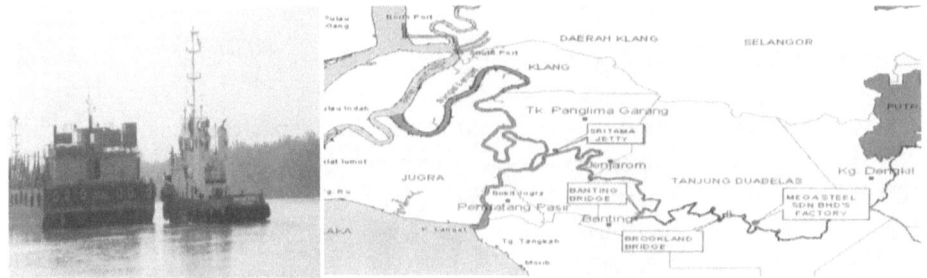

a. Barge and Tug convoy　　　　　b. Map of Langat River,

c. Vessels alongside

**Figure 3.33:** Langat River area, (JPS, 2008)

## 3.12.2 Langat River Baseline, Bathymetry and Profile

Described below is other information relating to channel, vessel and environment that is employed in the risk process. Lacking information about the distribution of transits during the year, or about the joint distribution of ship size and flag particular, environmental conditions become main derivative from probabilistic estimation. Langat river outlet splits into two directions as it approaches the sea thus creating Lumut Island in between. The north outlet enters the sea at the eastern side of Pulau Indah. The depth here varies from 2 to 6 meters. The south outlet enters the sea at Tanjung Tongkah towards Kuala Langat. The offshore depth at Kuala Langat is generally shallow ranging from 1 to 3 metres deep. The beaches along Kuala Langat are characterised by a presence of extensive mud flat stretching up to 200 metres from the high water mark. The mud flat consists of a mixture of fine sand and clays The bathymetry offshore the estuary is determined by constant activity of dredging due to the presence of the harbour. Bathymetry map (new edition 1998) shows that the depth at the mouth of the estuary is shallow ranging from 1 to 5 meters deep. While at mid channel depth increases drastically to 6 to15 metres. The two types of depth have been maintained by the port authority to ensure sea traffic viability plying the channel towards the Klang South Port (DOE, 2008).

## 3.12.3 Langat River Navigation

Kuala Sungai Langat was declared by marine department as a port. The port limits is from a point on the Southern coast of Carey Island at latitude 02° 49.8' North, Longitude 101° 20.0' East due Southernly to Latitude 02° 45.0' North, Longitude 101° 20.0' East thence due South Easternly to latitude 02° 43.5' North, Longitude 101° 23.0' East thence due North Easternly to D. Langat currently has approval for maximum of 5000t barge. Recent channel improvement for water deepening and widening work on Langat River follows PIANC standard, coastal engineering manual, Canadian Coast Guard manual. Langat River is currently maintaining minimum water way depth of 5.2 m with design bed level is based on lowest astronomical tide, from model simulation result. Figure 4.32 shows the Langat River tributary. Barge and tug of capacity 5000T and 2000T are currently plying this waterway at draft of 9 to 15 respectively. Collisions (including contact between two vessels and between a vessel and a fixed structure), causes of collision are linked to navigation system failure, mechanical failure and vessel motion failure. These factors and their interconnection are considered in PrHA towards use of the river for

transportation. Safety associated with small craft or other object or waste
signature is not taken into account (LUAS, 2009).

This section presents the data require from Langat river to implement
the model presented in chapter 4 and the result of implementation for IWT
safety and environmental risk deduced from the system engineering reliability
based approach. The result of the implementation is discussed in chapter
four. The most important environmental factors affecting the Langat River
are marine processes, particularly the tides and tidal currents. Tides in the
Malacca Straits have a mean spring range of 4.3 m. Tidal currents are bi—
directional, but set to the northwest, and average tidal current velocities range
from 1.5 m/sec to the northwest to less than 1.0 m/sec to the southeast. This
constant tidal reworking of bottom sediments in the shallow (120-180 m)
straits results in concentration of a large sand body. Those are characterized
by large asymmetrical sand waves along the bottom of the Malacca Straits.
Table 3.29 shows Langat River improvement activities.

**Table 3.29:** Past River improvement activities on Langat River (JPS, 2008)

| Date | Activities |
|---|---|
| Oct-05-Feb-06 | Data collection from River and Coastal Hydraulic Study |
| Jan-06 | Study for the navigation of 2000T Barge |
| May-June-06 | Megasteel Appointed Licensed Surveyor to Survey the River Corridor |
| Aug-06 | River Excavation for 2000T Barge |
| Oct-06 | Approval from Marine Department on Anchorage Points and Navigational Aids |
| Nov-06 | Trial Navigation of 2000T Barge Starts |
| Nov-06 | Intent for new River Straightening Alignment |

The reason for choosing examples of maintenance of Langat River for
the case study is because of recent environmental pressure emanating from
global warming, climate change, and population rise leading to congestion
in the city and point form pollutant release. Envisaged expansion in the area
further forecast potential rise of traffic in that area for a long time. Intolerable
system failure on complex and dynamic system like IWT has made decision,
policy and political will very hard. This make the need for use of scientific
system based model deduce from risk and high goal based objectives couple
with uncertainty analysis necessary for IWT development in Langat River.
This study look into the balancing system between safety, environment,
risk, uncertainty, social, economic, and technical requirement leading to

sustainable development of reliable and efficient IWT. Development of IWT involve assessment of channel and vessel size, maintenance dredging, soil, air and water quality, sustainability, risk and uncertainty. These factors has been disused in chapter two of this thesis.

The main risk contributing factors to collision are: power transmission failure, Navigation failure, Vessel motion failure, Human error, other external cause. For visibility, navigation is more risky at night than day time. The analysis follows generic assumption for evenly safe distribution during day and night. A review of risk assessment methodologies applicable to marine systems reiterate that the absence of data should not be used as an excuse for not taking an advantage of the added knowledge that risk assessment can provide on complex systems. Table 3.29 shows Langat River improvement activities done for channel. These remain part of risk consideration. Figure 3.34 shows Langat River channel width and dimension. Table 3.30 shows channel straight dimension data. Channel cross section, design has Straight Channel=53m, At Bend=64m.

**Table 3.30**: Straight Channel dimension

| Type | Channel dimension | |
|---|---|---|
| Straight | Basic Maneuvering Lane | 1.5B |
| | Addition for cross wind (less than 15 knots) | 0.0B |
| | Addition for cross current (negligible <0.2 knots) | 0.0B |
| | Addition for bank suction clearance | 1.0B |
| | Addition for aids to navigation (Excellent) | 0.0B |
| | Addition for cargo hazard (medium) | 0.0B |
| | Channel width for Inland Waterways, (B= Beam of the ship) | 2B=53 m |
| Bend | Channel width | 3B=64 m |

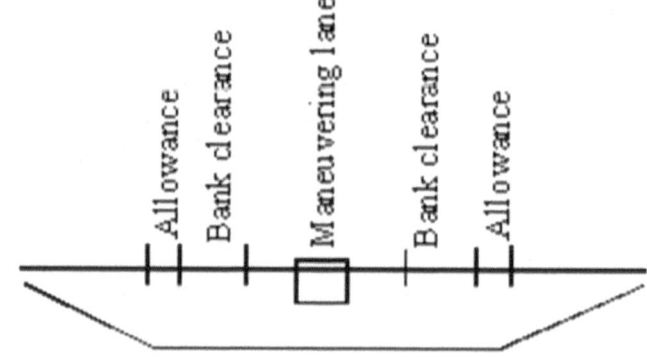

**Figure 3.34**: Channel width and depth dimension

## 3.12.4 Vessel Principal Dimension Data

Table 3.31 shows approach channel. Table 3.32 and 3.33 below shows the vessel principal dimension used for the analysis.

**Table 3.31:** Approach Channel

| | Approach channel | |
|---|---|---|
| Design parameter | Straight | Bend |
| | 98m | 120m |
| | 3-6m | 3-6m |
| Side slope | 10H:1V | 10H:1V |
| Estuarine 135.7km | North (44.2km) | South (9.9km) |

**Table 3.32:** Vessel dimension requirement (barge)

| Barge parameter | Tonnage | |
|---|---|---|
| | 2000 tons | 5000 tons |
| Length (m) | 67.3 | 76.2 |
| Beam (m) | 18.3 | 21.3 |
| Depth (m) | 3.7 | 4.9 |
| Draft (m) | 2.9 | 4 |

**Table 3.33:** Vessel dimension requirement (tugs)

| Tugs Parameter | Tonnage | |
|---|---|---|
| | 2000 tons | 5000 tons |
| Length (m) | 23.8 | 23.8 |
| Beam (m) | 7.8 | 7.8 |
| Depth (m) | 3.5 | 3.5 |
| Draft (m) | 2.8 | 2.8 |
| Horse Power (hp) | 1200 | 1200 |

## 3.12.5 Langat River Traffic Data for Accident Frequency Modeling

Vessel movement and port call usually consists of two transits: i) One into the port and ii) One out of the port. Safe transit data consider the same barge type and size (Refer to Table 3.34, 3.35 and 3.36). Table 3.37 shows the chronology of accident along Sungai Langat due to inland navigation activities.

**Table 3.34**: Tug and barge activities along Langat River for 2008

| Jetty | 3 nos |
|---|---|
| Daily | 9 times |
| Weekly | 63 times |
| Monthly | 252 times |
| Annually (t) | 3024 times |

**Table 3.35**: Number of ships passing

| Total Number of Barge | Time | Traffic |
|---|---|---|
| 12 | Everyday (24 Hrs) | |
| 6 | Every (4 Hrs) | Incoming |
| 6 | Every (4 Hrs) | Outgoing |

**Table 3.36**: Common to traffic

| All speed (s) | 2~3 knots |
|---|---|
| All traffic | Single way traffic |
| River improvement programme | Proposed locations for Lay-bys |

### 3.12.6 Langat River Accident Data

Figure 3.35 shows accident at Langat River. Table 4.19 shows the chronology of accident along Sungai Langat due to inland navigation activities.

**Figure 3.35**: Accident at Langat River

**Table 3.37**: Chronology of accident along Langat River
due to inland navigation activities

| No | Date | Time | Case | Impact description |
|---|---|---|---|---|
| 1 | 12.01.08 | 8.30pm | Hit | 2 nos. of concrete pile have a scratched mark, 5 number of concrete pile was broken. |
| 2 | 25.02.08 | night | Hit | The protection pipe nearby jetty was damaged & missing. 3 numbers. of welded pipe were missing. |
| 3 | 08.03.08 | 930pm | Hit/ Tug | The protection pipe nearby jetty was damaged & missing. 3 numbers. of welded pipe missing/bended I—beam |
| 4 | 07.04.08 | 3.30pm | And | Temporary jetty at Pier 9a. |
| 5 | 07.05.08 | 5.00pm | vessel | Temporary jetty at Pier 9a. |
| 6 | 09.05.08 | 8.30am | | Temporary jetty at Pier 9a. |
| 7 | 17.06.08 | 4.00am | | I-Beam that used for Gen-Set was bended. Workers jetty was badly damage. |
| 8 | 28.07.08 | 10.30am | | Bended I-Beam. |
| 9 | 23.08.08 | 8.30am | | Protection pipe at Pier 9a. |
| 10 | 18.09.08 | 6.30am | | Temporary jetty at Pier 9a. Welding set & crane were collapsed & drown into the river. |

## 3.12.7 Environmental Data

The main environmental factors are describing in Figure 3.36 a, b and c show the environmental parameters considered in the risk process. In Figure 3.36a, Tide movement and current EB, Green to red: low to high speed.

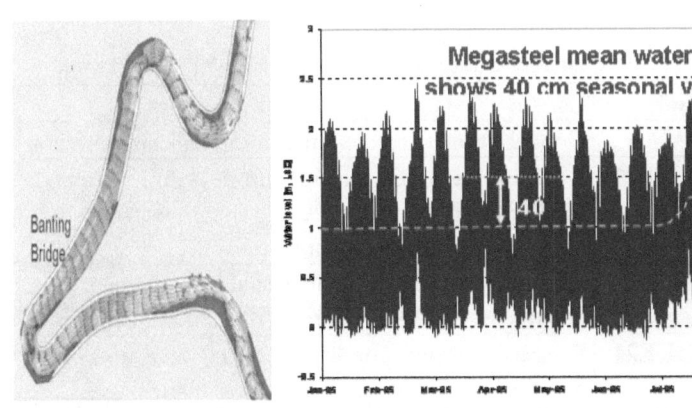

a. Tide movement                    b. Water level Mean,

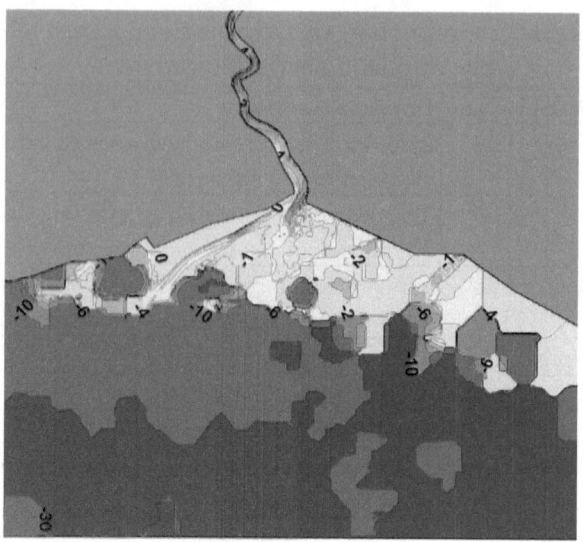

c. Coastal and current EB

**Figure 3.36:** Langat River Environmental Information

In Figure 3.36b, Water level Mean, water level = 40cm seasonal variation, Existing coastal environmental current In Figure 3.36c, Coastal current, Avg. Speed in Spring tide 0.4—1.2 m/s, Avg. Speed in Neap 0.2—1.0 m/s. Table 3.38 shows previous model used for Langat River. Identification of accident scenario that is significant to risk contribution considers the use of holistic qualitative risk assessment that high level objective derived from system functionality and standards matching.

**Table 3.38:** Past River Models on Langat River

| Model | Tool | Use for |
|-------|------|---------|
| Rainfall-Runoff | NAM | Contribution of catchments runoff to Langat river |
| One-Dimensional River | MIKE11 | Establish baseline condition of tide, salinity, flood level of the Langat river Assess the impacts of navigational improvement plan |
| Two-Dimensional Curvilinear Grid | MIKE21 C | assess the impacts of navigational improvement plan on erosion/deposition pattern of the Langat river |
| Two-Dimensional Rectangular Grid Nested | MIKE21 | Establish baseline condition of tide, wave, erosion/ deposition pattern for Langat river mouth assess the impacts of proposed navigation improvement plan |

A point on the coast of Kampung Tanjung Tongkah, near the estuary of
Sungai Langat at Latitude 02° 47.25' North, Longitude 101° 25.0' East and
thence following along the Eastern bank of the river Sungai Langat until
reaching a point on an imaginary line at Latitude 02° 50.35' North, Longitude
101° 24.8' East point on the Western bank. Hence due West towards a of the
river Sungai Langat at Latitude 02° 50.35' North, Longitude 101° 24.65' East,
returning downstream along the river bank until the estuary of Sungai River.

## 3.13 Channel Complexity and Uncertainty Analysis

Issues associated with uncertainty are influences on recovery process,
test of new advancements, and influence on policy, system, hangs over time,
services and resources. Dealing with uncertainty requires quantifying critical
uncertainty and making sustainability balance between safety, economics
and environmental sustainability. Model on weather and human reliability
assessment, and expert judgment as well as simulation could help perfect the
reliability on safety and environmental risk study for IWT. Uncertainty is
necessary because of highly variable nature of elements and properties involved
with the situation. It is required to simulate extreme condition and model by
using combination of mathematical modeling and stochastic techniques.

Measure of uncertainty could target the: i) Historical risk data assessment
(people > other species), ii) Mitigation of consequence (immediate threat
> delayed threats), iii) Prefer and no option choice (treatable > untreatable),
iv) Community participation and expert rating (more certain > less certain)
Emergency response, monitoring and information dissemination v) Risk
areas and assessment: using historical data and statistics that include all
factors. Public health (people greater than other species), vi) Mitigation to
risk assessment and risk areas; vii) Prefer and no option choice: compensation
strategies, viii) Panel of expert, x) Uncertainty (More certain less than certain),
xi) Community participation, xii) Emergency response, xiii) Other technique
to increase reliability and reduce uncertainties as much as possible is taken into
consideration in the quantitative analysis.

**Table 3.39:** Quantitative risk and reliability tools

| Quantitative tools | Application |
|---|---|
| Frequency, Consequence and ALARP Analysis | **Involve analysis of causal factor and impact of accident** |
| Failure Modes and Effects Analysis (FMEA) | Use to analyse the components (equipment) failure modes and the impacts on the surrounding components and the system |

| Fault Tree Analysis (FTA) | Use to analyse combinations of equipment failures and human errors that can result in an accident |
|---|---|
| Event Tree Analysis (ETA) | Use to analyse various consequences of events, both failures and successes that can lead to an accident. |
| Technique for Human Performance Reliability Prediction (THERP) | Use to analyze human error |

Table 3.39 shows quantitative risk and reliability modeling tool. Techniques are employed to analyse uncertainty analysis. The risk modeling utilizes fault tree and event tree analysis of the important risk contributing factors to the accident scenarios, in this case collision. Other technique to increase reliability and reduce uncertainties as much as possible is taken into consideration in the quantitative analysis are described in the following section.

### 3.13.1 Failure Mode Effect Analysis

Figure 3.37 shows FMEA process for risk analysis. The FMEA analysis FMEA involves account for the following action and check in risk analysis, description of unit, and description of failure, effect of failure, failure rate, severity ranking, risk reduction measure and recommendation.

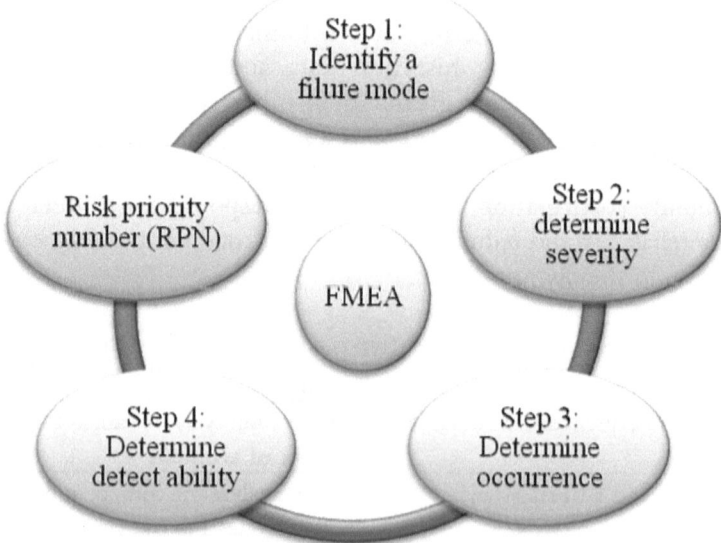

**Figure 3.37:** Processes of FMEA

Failure modes effect analysis (FMEA) is the least complex risk assessment tool. Analyzing FMEA involves (Moderras, 1993), below are the FMAEA steps whose result is given in table in Chapter 4:

i.   Identify failure mode, list key process steps in column for the highest ranked items risk matrix
ii.  Determine severity, list the potential failure mode for each process step
iii. Determine delectability, list the effects of the failure mode, this include stakeholder feeling
iv.  Risk priority rate how severe this effect is with 1 being not severe at all and 10 being extremely severe, identify the causes of the failure mode effect and rank it. Identify the controls in place to detect the issue and rank its effectiveness in the detection column.

Severity, occurrence, and detection numbers is multiplied and store this value in the risk priority number (RPN) column. This is the key number that will be used to identify where risk analysis should focus first. If there is outcome with severity of 10 (very severe), occurrence of 10 (happens all the time), and detection of 10 (cannot detect it) Risk Priority Number (RPN) is 1000, that indicate sign of serious issue. Sort by RPN number and identify most critical issues. The team must decide where to focus first and analysed accordingly. Assign specific actions with responsible persons. Once actions have been completed, score the occurrence and detection is computed.

### 3.13.2 Fault Tree Analysis (FTA)

FTA is a logical and diagrammatic method to evaluating the occurrence probability of an accident resulting from sequences of faults and failure events. FTA is useful analysis of mode of occurrence of an accident using logic gate. It is used to deduce the probability of top event from the failure probabilities occurrence of system components. Fault tree analysis are generally performed graphically using logic AND and OR gates. It has Top Event from which other contributing factors tree down to prediction probability number for the event. The "AND" Gate describes the logical operation where the output event occurs if all input events contribute to the accident occurrence. An "OR" Gate defines the situation whereby the output event will exist if at least one of the input events is present. Here are no restrictions on the number of inputs to each gate. Figure 3.38a and b show the symbols use for FTA construction. FTA involves five steps. Define the undesired event to study; Obtain an understanding of the system, Construct the fault tree, Evaluate the fault tree and Control the hazards identified.

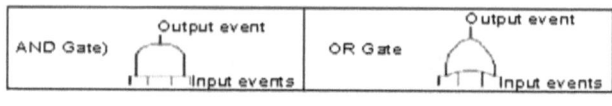

**a**: AND and OR Logic Gates

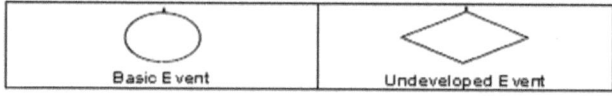

**b**: Typical Primary Events

**Figure 3.38:** FTA Logic gate

The primary events of the fault tree are those events for one reason or another which have not been further developed. The basic event describes a basic initiating fault event that requires no further development. The diamond describes a specific fault event that that do need further development of accident, either because the event is of insufficient consequence or the necessary information is unavailable (See Figure 3.38). The occurrence probability of the top event of a fault tree can be calculated when the probabilities of the occurrence of basic fault events that are known. This can be obtained by first calculating the occurrence probability of the resultant (i.e., output) fault events of intermediate lower logic gates like AND and OR. The occurrence probability of the top event of collision can be compared to from risk acceptability table. The probability of occurrence of the AND gate output fault event is expressed by:

$$P(x_o) = \prod_{i=1}^{k} P(x_1)$$
Eq. 3.108

Where: $P(x_o)$ is the probability of occurrence of the AND gate output fault event, $x_o$.$n$ is the number of AND gate input fault events. $P(x_i)$ is the occurrence probability of AND gate input fault event $x_i$; for $i = 1, 2, 3, \ldots, n$

Similarly, the probability of occurrence of the OR gate output fault event is given by:

$$P(x_o) = \prod_{i=1}^{n} \{1 - P(y_1)\}$$
Eq. 3.109

Where: $P(y_0)$ is the probability of occurrence of the OR gate output fault event, $y_0$.

$k$ is the number of OR gate input fault events. $P(y_i)$ is the occurrence probability of

OR gate input fault tree event $y_i$; for $i = 1, 2, 3, \ldots, k$

### 3.13.3 Event Tree Analysis (ETA)

Event tree analysis is analysed by binary and discreet logic, in which an event either ON or OFF denote occurrence or an indication that a component of it has not failed. It is valuable in analyzing the consequences arising from a failure or undesired event. An event tree begins with an initiating event, such as a component failure, increase in temperature and pressure or a release of a hazardous substance. The consequences of the event are followed through a series of possible paths. Each path is assigned a probability of occurrence from where the probability of the various possible outcomes can be calculated. Event tree analysis process involves as in Figure 3.39a. The process start with initiating event following by number of pivotal events and end state is analysis with assigned probability according to percentage contribution to the event (Figure 3.3b).

a: Event Tree Analysis process

ETA involves; Define the system or area of interest to specify and clearly define the boundaries of the system or area for which event tree analyses will be performed. ii) Identify the initiating events of interest. iii) Identify lines of assurance and physical phenomena. Identify the various safeguards (lines of assurance) that will help mitigate the consequences of the initiating event. These lines of assurance include both engineered systems and human actions. Also, identify physical phenomena, such as ignition or meteorological conditions that will affect the outcome of the initiating event.

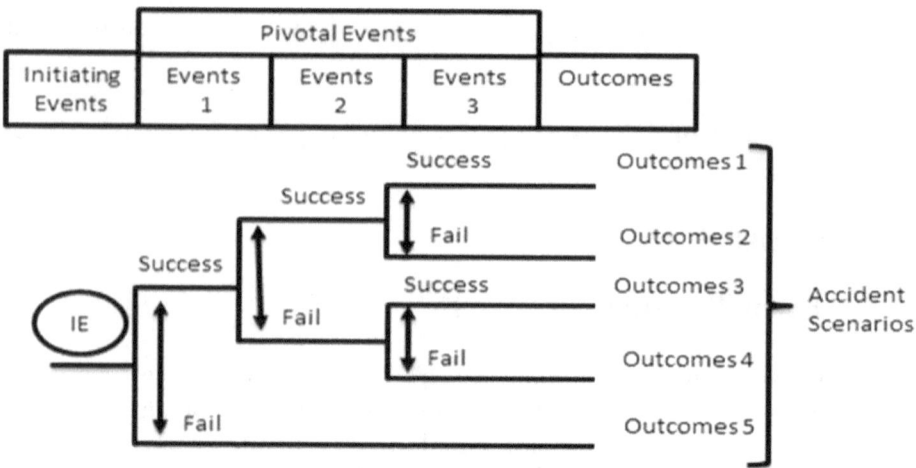

b: Event Tree Analysis Outcome

**Figure 3.39:** Event Tree Analysis

For subsystem level analysis propulsion failure, human factor, loss of mooring function is considered using RELEX software for fault tree and event tree analysis. FMEA and FMCEA can also be used for further analysis of subsystem. Risk result can be compared with the risk acceptability criteria where SOD is determine for most serious risk category. The consequence analysis for subsystem contributing factors should be performed using event tree analysis ETA while frequency for subsystem contributing factors can be modeled by FTA. Risk models can be developed for each accident scenario where ETA and FTA is constructed according to the risk models and considering data regarding accident statistics, damage statistics, fleet statistics, analytical modeling and expert opinion.

### 3.13.4 Propulsion Failure Analysis

Scenario and components risk analysis like propulsion failures should start with system definition. The definition includes system hazard analysis, preliminary safety assessment and system safety assessment after system functionality and standard matching. Figure 3.40a and b gives a simple description of propulsion system.

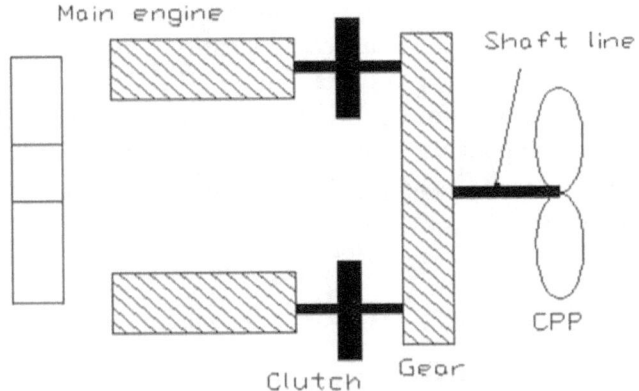

**Figure 3.40a:** Propulsion system descriptions

Prediction of the reliability of the propulsion and steering gear of a ship is done by incorporating a qualitative system description with a quantitative description. The propulsion, the steering and the electrical power network are combined into one to describe the entire system. Propulsion failure analysis could also include maintainability to reduce the failure rate. The propeller is key part of engine motion mechanic; propeller failure can lead to immobilization of whole propulsion system, navigation system impairness and consequential uncontrollability of ship motion.

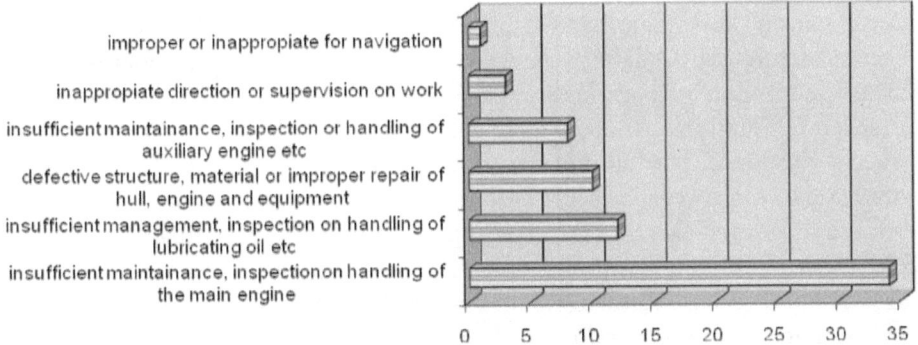

**Figure 3.40b:** Cause of machinery failure

Figure 3.40 depict major cause of machineries failure leading to propulsion failure. The propeller is design to transform rotational energy transmitted through propeller shaft into a pressure difference over the propeller blades which accelerate and maintain the speed of the vessel. The analysis towards selection of propulsion system begins with system description. This is followed by matching the functional and regulatory requirement. Functional requirement of propulsion system include operational components: normal speed, maneuvering operation condition, life period of operation, period for maintenance, period in harbour, full power sailing time, period for port operation, crew, maintenance planning, system components: main diesel engine, clutch, gear, shaft line, controllable pitch propeller, spare parts tool, s and ystem connection: Series and parallel probability model can be incorporated.

$$P_{sf} = 1 - (1 - P_1) + (1 - P_2) + \ldots + (1 - P_n) = \prod_{i=1}^{n} P_i \text{ (parallel structural part)} \quad \text{Eq. 3.110}$$

$$P_{pf} = 1 + 1 - P_2 \ldots + P_n = \prod_{i=1}^{n} (1 - P_i) \ldots \text{(For series part)} \quad \text{Eq. 3.111}$$

A typical system selection could be a diesel electric unit following the above describes system components. Operation may demand that that two diesel engine could be used when high power is required and they gear can be decoupled is maximum power.

## 3.13.5 Human Reliability Analysis

The dynamic distributive condition, long incumbent period and complexity of marine system with limited oversight make the process of identification and addressing human and organizational error difficult. This makes human reliability analysis at best to be modeled independently. Expert judgments give evaluation of the likelihood that failures would occur in specific situations can be use to quantify human reliability input in risk process Evaluation and comparing baseline scenario to a set of scenarios of interest like tug or barge escort case in waterway can be considered. Human errors and human factors are recommended to be studied separately. Figure 3.41 shown components of elements require to be considered for either case. THERP is the best known, most frequently applied technique for human performance reliability prediction. It is a method for predicting human error rates and for evaluating the degradation of a human-machine system likely to be caused by human errors in association with factors such as equipment reliability, procedures, and other factors. THERP uses performance-shaping factors to make judgments about particular situations. In some cases, however, it may be difficult to accommodate all of the factors that are considered in the system. THERP has the advantage of simplicity but it does not account for a dependency of human performance reliability with time.

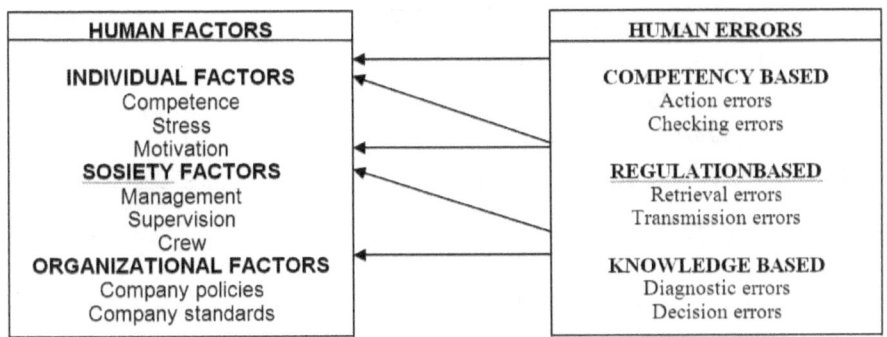

**Figure 3.41:** Human Factors vs. Human Errors

THERP processes include: defining the system or process. His is followed by identify and list all the human operations performed and their relationships to the system or process tasks and functions. Predict error rates for each human operation or group of operations. Then, determining the effect of human errors on the system or process, including the consequences of the error not being detected and develop and recommend changes that will reduce the system or process failure rate. The recommended changes can be developed using sensitivity analyses. Human reliability analysis caters for uncertainty estimation. It is important for this to be carried out separately. THERP is use model criticality and stochastic estimation human reliability. This can be done using questionnaire analysis or the THERP and probabilistic relation. THERP is represented by:

$$P_{EA} = HEP_{EA} .k \sum_{k=1}^{m} PSF_K .W_K + C$$

Eq. 3.112

$P_{EA}$ is Probability of an error for a specific action, basic (nominal) probability of a specific action, $PSF_k$ is Numerical value of Kth performance shaping factor, $W_k$ is Weight of $PSF_k$ (Numerical constant), C is Numerical constant, m is Number of PSFs. Table 3.40 shows risk consequence acceptability criteria in maritime industry.

**Table 3.40:** Risk acceptability criteria for consequence analysis

| Quantification | Serenity | Occurrence | Detection | RPN |
|---|---|---|---|---|
| current failure that can result to death failure, performance of mission | Catastrophic (10) | 2 | 10 | 200 |
| failure leading to degradation beyond accountable limit and causing hazard | Critical (7) | 2 | 5 | 70 |

| controllable failure leading to degradation beyond acceptable limit | Major (5) | 4 | 2 | 20 |
| nuisance failure that do not degrade system overall performance beyond acceptable limit | Minor (1) | 6 | 2 | 12 |

$$RPN = S \times O \times D \qquad\qquad \text{Eq. 3.113}$$

Right rating for SOD can be choose from frequency, consequence and detection risk acceptability table

### 3.13.6 Criticality Analysis

HAZID and HAZOP involve identifying hazard that can result to system failure. HAZOP assessments involve combination of causes and measure that deal with identification of hazards to prevent accident. HAZID is associated with propeller failure that could be avoided could use guideword and "what if" example that can facilitate the assessment. FMEA and FMECA involves inductive method to determine equipment functionality, failure mode, cause of failure consequence, impact, reliability, safety, cost and quality of system components.

FMEA uses standardized form that account for qualitative measure of failure rate and severity level. FMECA is advance FMEA where criticality of failure is quantitatively accounted for the reliability and maintainability of the system. Analysis of FMCEA provides answer to the following (Moderas 1993): i) system description, possible failure and mode of failure, ii) cause of failure and effect of failure, iii) grading effects according to frequency, iv) severity of consequence and reliability data, v) specify and assessment method for detection status and vi) mitigation of unwanted effects towards reduction and elimination. The criticality quantification can be done by applying:

$$C = \sum E_i L_i T \qquad\qquad \text{Eq. 3.114}$$

Where: E= Failure consequence probability of failure mode i, L = occurrence likelihood of failure mode i, Number of failure mode of the item which fall under particular severity class, T=Duration of application of mission place.

## 3.13.7 Channel Visibility Analysis

Various channel components enter channel complexity regime. These can be visibility weather, squat, bridge, river bent, human reliability. It is important to account for each of them in channel design work. Poor visibility might be expected to increase the risk of groundings and collisions. The increase in accident risk due to poor visibility is most likely to be more significant than the change associated with high wind. A visibility model from Dover waterway studies concluded with the following (Thomas, 2003).

Fog Collision Risk Index (FCRI) = $(P_1 + VI_1 + P_2 + VI_2 + P_3 . VI_3)$   Eq. 3.115

Where: $P_k$ = Probability of collision per million encounters, $VI_k$ = Fraction of time that the visibility is in the range k, K = Visibility range: clear ( > 4km), Mist / Fog (200m—4km), Tick/dense (less than 200m). The empirically derived means to determine the relationship between accident risk, channel complexity parameters and VTS is represented as followed:

$$R = - 0.37231 - 3.5297C + 16.3277N + 0.2285L - 0.0004W + 0.01212H + 0.0004M$$   Eq. 3.116

Where: Predicted VTS = consequence for 100,000 transit, C = 1 for an open approach area and 0 otherwise, N = 1 for a constricted waterway and 0 otherwise, L = length of the traffic route in statute miles, W = average waterway and channel width in yards, H = sum of total degrees of course changes along the traffic route, M = number of vessels in the waterway divided by L.

Barge movement creates very low wave height and thus will have insignificant impact on river bank erosion. Speed limit can be imposed by authorities for wave height and loading complexity. Consideration for river tributary and bent also enter channel complexity work.

## 3.13.8 Squat Analysis

Navigation of vessel in shallow water at a hull displacement cause vertical sinking, or squat, this is as a result of a pressure drop beneath ship hull. To avoid ship groundings with possible severe economic and environmental consequences, the relevant governmental, port, maritime agencies and organizations need a reliable method of predicting ship squat (Eryuzlu, 1994). Analysis of squat and channel clearance is based on the physical characteristics of a channel and the ships that travel through it. It can be used to issue appropriate regulations regarding vessel size, speed and to plan channel dredging operations. Squat calculation can be done using the following Huska formula.

$$\frac{\Delta_t}{t} = C_O \frac{C_B b}{l_{pp}} \frac{F^2}{\sqrt{1 - F^2 nh}}$$

Eq. 3.117

Alternative formula recommended by PIANC/IAPH is:

$$\text{Squat (m)} = \frac{2.4 . \dfrac{\nabla}{L_{PP}^2} . \nabla \dfrac{F^2}{\sqrt{1 - \dfrac{2}{nb}}}}{}$$

Eq. 3.118

Where, $\nabla$ is the displacement = CLBT, C is block coefficient, L is length perpendicular, B is the beam of the ship and T is the draft of the ship and F is the Froude depth number.

### 3.13.9 Mooring and Berthing Energy

Mooring and berthing risk are classified as of high, medium or low can be determined based on the following site specific parameters like wind, current, hydrodynamic effects of passing vessels, change in vessel draft and the maximum wind and maximum current. Table 3.31 shows mooring and berthing classification under SOLAS (IMO, 2006).

**Table 3.41**: Mooring and berthing risk classification (Harwish, 1992)

| Risk Classification | Wind, $(V_W)$ (knots) | Current, $(V_c)$ (knots) | Passing Vessel Effects | Change in Draft (ft.) |
|---|---|---|---|---|
| High | >50 | >1.5 | Yes | >8 |
| Moderate | 30 to 50 | 1.0 to 1.5 | No | 6 to 8 |
| Low | <30 | <1.0 | No | <6 |

The berthing velocity can be obtained from actual measurements or relevant existing historical information. When the actual measured velocity is not available, the PIANC or others standard can be adopted to determine the required velocity value. The berthing energy is used to determine the most suitable fender for the berthing facilities. For Langat River ship range between 1000-10000 tons displacement barges and tug is considered. When the ship is stopped by the fender, the momentum of the entrained water continues to push against the ship and this effectively increases its overall mass. The berthing energy is calculated from the following relation (Carr, 1952):

$$E = 0.5[\frac{(Md_1 \times Cm_1) \times (Md_2 \times Cm_1)}{(Md_1 \times Cm_1) + (Md_2 \times Cm_1)}] \times V^2 \times Ce \qquad \text{Eq. 3.120}$$

Where; the mass of specified water is referring to Added Seawater Mass; the added seawater influence coefficient is Cm. Winds, currents, and waves have loading on moored vessels. Dynamic loads are due to waves. Static loads due to wind and current are separated into longitudinal load, lateral load, and yaw moment. Loads on moored vessels due to wind result primarily from drag. The lateral and longitudinal wind load can be determined from.

$$F_{yw} = \frac{1}{2}\rho_a V_w^2 A_y C_{DW} sin\theta_w \qquad \text{Eq. 3.121}$$

$$F_{xw} = -\frac{1}{2}\rho_a V_w^2 A_x C_{DW} cos\theta_w \qquad \text{Eq. 3.122}$$

Lateral forces are dominated by form drag. Form drag depends on the ratio of vessel draft to water dept. When the water depth decreases, current flows around rather than underneath the vessel, longitudinal forces due to current are caused by form drag, friction drag, and propeller drag. The lateral and longitudinal current load can be determined from:

$$F_{yc} = \frac{1}{2}\rho_w V_c^2 L_{wL} T C_{yc} sin\theta_c \qquad \text{Eq. 3.123}$$

$$F_{xc} = F_{x\,form} + F_{x\,fricion} \qquad \text{Eq. 3.124}$$

Forces generated by passing vessels are due to pressure gradients associated with the flow pattern. These pressure gradients cause the moored vessel to sway, surge, and yaw. Passing vessel analysis should be conducted when all of the following conditions exist that can be determine from equation 3.88. Other consideration is that if L £ 2B, passing vessel loads should be considered. L and B are shown in Figure 3.42 in units of feet. V is defined as the speed of vessel minus the current velocity, when traveling with the current, or the speed of vessel plus the current velocity, when traveling against the current.

$$V_{crit} = 1.5 + \frac{L - 2B}{500 - 2B}4.5 \quad \text{(Knots)} \qquad \text{Eq. 3.125}$$

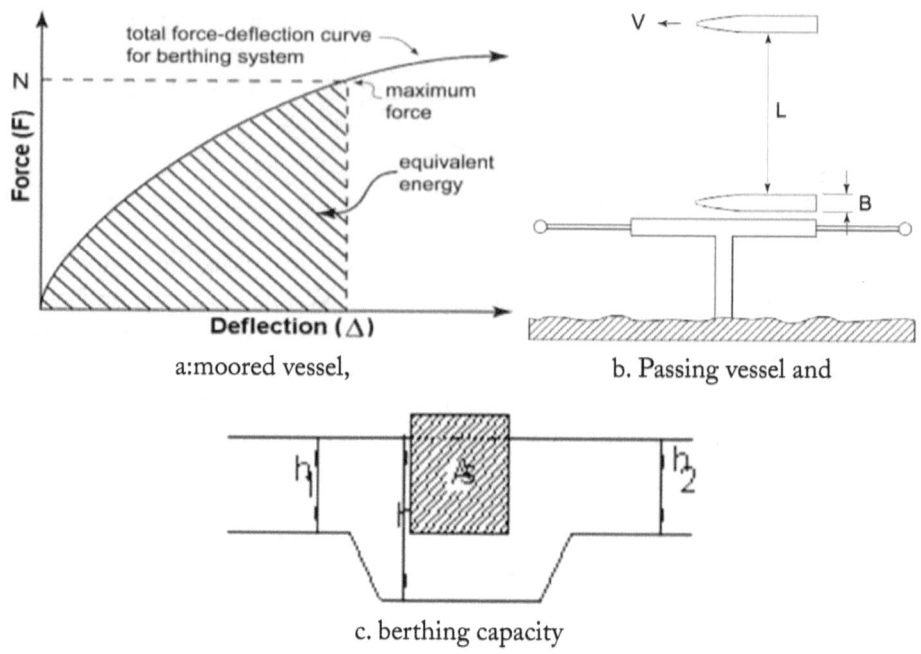

a:moored vessel,

b. Passing vessel and

c. berthing capacity

**Figure 3.42:** Mooring requirement

When such conditions exist, the surge and sway forces and the yaw moment acting on the moored vessel should be assumed to be at minimum. If the demands from such evaluation are greater than 75% of the mooring system capacity (breaking strength of mooring lines), a more sophisticated dynamic analysis is required. Passing vessel size should not be greater than 25,000 dwt. Distance L is 500 feet or less, vessel speed V is greater than $V_{crit}$ (DON, 1984). The fender system should be designed to absorb the berthing energy. The berthing energy capacity should be calculated as the area under the force deflection curve for the combined structure and fender system. Fender piles may be included in the lateral analysis to establish the total force deflection curve for the berthing system. Load deflection curves for other fender types can be obtained from manufacturer's data. The condition of fenders should be taken into account when performing the analysis. Fender is use as a leading anti-collision device for berthing facilities. Fender is used as a protective medium. The advantages of using fender include easy and fast to deploy, very low reaction and hull pressure, performance adjustable by varying initial pressure suitable for areas with large or small tides low maintenance.

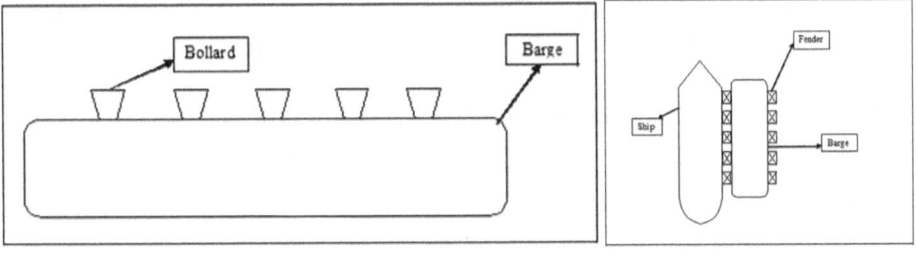

a. Bollard                              b. Fender

**Figure 3.43:** Berthing facilities for the facilities

Figure 3.43 illustrated the arrangement of the bollard on floating pier. Bollard is use on quay for mooring. In this study, it is used to fasten the floating pier to hold vessels steady by ropes against the turbulence during berthing operation. Mooring hooks set in recesses in the walls provide an alternative anchorage against surging.

### 3.13.9.1 Berthing Energy

In most cases, the actual values of the ships, displacement tonnage, berthing velocity, added mass coefficient and eccentricity coefficient are the important parameter that are used to calculate the actual berthing energy. Figure 3.44 showing the modeling process. Berthing energy is determined by analyzing water displacement of ship, added mass coefficient, berthing velocity, incorporation of eccentricity coefficient and deduction of berthing energy.

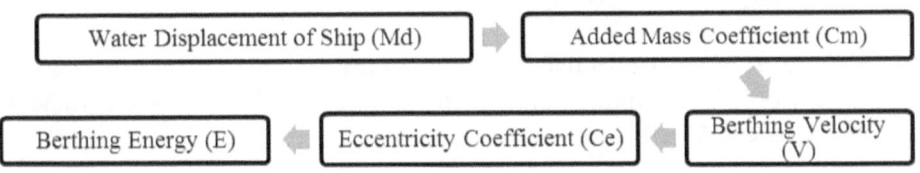

**Figure 3.44:** Berthing energy process

Water Displacement of Ship range from 1000-10000 tons in water displacement is choosing according to size of the ships that transit or berth Langat River. Added Mass Coefficient involve empirical formula employ for the added mass coefficient (Cm)

$$Cm = 1 + \frac{\pi D^2 L \rho}{4Md} \qquad\qquad \text{Eq.3.126}$$

Where: Cm =Added Mass Coefficient, D =Draft (m), L= Ship Length (m), $\rho$ = Seawater, Density (t/m³), Md= Water Displacement of ship.

In Berthing Velocity Empirical formula used for berthing speed (m/sec) is

$$v = \frac{1 - \frac{1}{\log(Ms)}}{\sqrt[3]{\left(\frac{Ms}{1000}\right)}}$$
Eq.3.127

Where: $v$ = Berthing Speed (m/sec), $M_s$ = Loaded Water Displacement of the ship (tons),

Eccentricity Coefficient: Empirical formula for the Eccentricity Coefficient (Ce):

$$Ce = \frac{1}{1 + \left(\frac{l}{r}\right)^2}$$
Eq.3.128

Where: = Gyration radius of ship against axial of enter of gravity on horizontal plane.

I = Project of the distance between the center of gravity and berthing point on dock direction. The impacting energy calculation is subject to the ships berthing method which can be defined as Ship-To-Ship Berthing energy empirical formula:

$$E = 0.5 \left[ \frac{(Md_1 \times Cm_1) \times (Md_2 \times Cm_1)}{(Md_1 \times Cm_1) + (Md_2 \times Cm_1)} \right] \times V^2 \times Ce$$
Eq.3.129

Where: E =Vessel effective berthing energy (ton.m), Md=Displacement Tonnage (ton), V=Berthing Velocity (m/s), Cm= Added Mass Coefficient, Ce= Eccentricity Coefficient

### 3.13.9.2 Mooring System

The US Navy moorings are classified as either fleet moorings or fixed moorings. A fleet mooring consists of structural elements, temporarily fixed in position, to which a vessel is moored. These structural elements include anchors, ground legs, a riser chain, a buoy, and other mooring hardware. Lines and appurtenances provided by vessels are not a part of the fleet mooring. Environmental loads which are winds, currents, and waves produce loads on moored vessels are considered. Static wind and current loads are discussed in detail below. Static loads due to wind and current are separated into longitudinal load, lateral load, and yaw moment. Flow mechanisms which influence these loads include friction drag, form drag, circulation forces, and proximity effects. Figure 3.45 shows the procedure to determine the barge mooring system. Mooring system is analised first, by investigation of mooring site, followed by assessment of vessel to be moored to determine mooring configuration and load, followed by environmental loading and comparison with safety compliance.

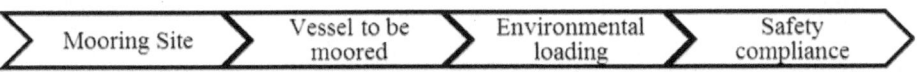

**Figure 3.45:** Mooring risk process

Lateral wind load on barge is determined using the following:

$$F_{yw} = \frac{1}{2}\rho_a V_w^2 A_y C_{DW} \sin\theta_w$$

Eq.3.130

Where: $F_{yw}$ = lateral wind load, in pounds, $\rho_a$ = mass density of air = 0.00237 slugs per cubic foot at 68°F, $V_w$ = wind velocity, in feet per second, $A_Y$ = lateral projected area of barge, in square feet, $C_{DW}$ = wind-force drag coefficient, $\theta_w$ = wind angle.

Longitudinal wind load on barge is determined using following:

$$F_{xw} = -\frac{1}{2}\rho_a V_w^2 A_x C_{DW} \cos\theta_w$$

Eq.3.131

Where: $F_{xw}$ = Longitudinal wind load, in pounds, $\rho_a$ = Mass density of air = 0.00237 slugs per cubic foot at 68°F, $V_w$ = Wind velocity, in feet per second, $A_x$ = Longitudinal projected area of barge, in square feet, $C_{DW}$ = Wind-force drag coefficient.

Wind yaw moment is computed using the following:

$$M_{xyw} = F_{yw}e_w$$

Eq.3.132

Where: $M_{xyw}$ = wind yaw moment, in foot-pounds, $F_{yw}$ = lateral wind load, in pounds, $e_w$ = eccentricity of $F_{yw}$, in feet, $e_w$ = $L[\frac{3.125}{100} - 0.0014(\theta_w - 90°)]$ for $0 \leq \theta_w \leq 180°$,

$$e_w = \frac{L[\frac{3.125}{100} + 0.0014(\theta_w - 270°)]}{} \text{ for } 180° < \theta_w \leq 360°, = \text{length of barge.}$$

Current loads developed on moored vessels result from form drag, friction drag, and propeller drag. Lateral forces are dominated by form drag. Form drag is dependent upon the ratio of vessel draft to water depth: as the water depth decreases, current flows around rather than underneath the vessel. Longitudinal forces due to current are caused by form drag, friction drag, and propeller drag.

Lateral current load is determined from the following equation:

$$F_{yc} = \frac{1}{2}\rho_w V_c^2 L_{wL} T C_{yc} sin\theta_c \qquad\qquad \text{Eq. 3.133}$$

Where: $F_{YC}$ = lateral current load, in pounds, $\rho_w$ = mass density of water = 2 slugs per cubic foot for sea water, $V_c$ = current velocity, in feet per second, $L_{wL}$ = vessel waterline length, in feet, T= vessel draft, in feet, $C_{yc}$ = lateral current-force drag coefficient, $\theta$ =current angle

The lateral current-force drag coefficient is given by:

$$C_{yc} = C_{yc}|oo + (C_{yc}|1 - C_{yc}|oo)e^{-k(\frac{wd}{T}-1)} \qquad\qquad \text{Eq.3.134}$$

Where: $C_{yc}$ =Lateral current-force drag coefficient, $C_{yc}|oo$ = Limiting value of lateral current-force drag coefficient for large values, of $\frac{wd}{T}$ .$C_{yc}|1$=Limiting value of lateral current-force drag coefficient for $\frac{wd}{T}$=1, $e$ =2.718, k=Coefficient, wd=Water depth, in feet, T=Vessel draft, in feet

Longitudinal current load procedures are taken from Cox (1982). Longitudinal current load is determined using the following equation:

$$F_{xc} = F_{x\,form} + F_{x\,friction} + F_{x\,prop} \qquad\qquad \text{Eq.3.135}$$

Where: $F_{xc}$ =Total longitudinal current load, $F_{x\,form}$ =Longitudinal current load due to form drag $F_{x\,friction}$ = Longitudinal current load due to skin friction drag, $F_{x\,prop}$ = Longitudinal current load due to propeller drag

Form drag is given by the following equation:

$$F_{x\,form} = -\frac{1}{2}\rho_w V_c^2 BTC_{xcb} cos\theta_c \qquad\qquad \text{Eq.3.136}$$

Where: $F_{x\,form}$ = Longitudinal current load due to form drag, $\rho_w$ = Mass density of water = 2 slugs per cubic foot for sea water, $V_c$=Average current speed, in feet per second, B=Vessel beam, in feet, T=Vessel draft, in feet, $C_{xcb}$ = Longitudinal current form-drag coefficient = 0.1, θ =Current angle. Skin friction drag is given by the following equation:

$$F_{x\,friction} = -\frac{1}{2}\rho_w V_c^2 SC_{xca}\cos\theta_c \qquad\qquad \text{Eq.3.137}$$

Where: $F_{x\,friction}$ = Longitudinal current load due to skin friction, $\rho_w$ = Mass density of water = 2 slugs per cubic feet for sea water, Vc=Average current speed, in feet per second, S=Wetted surface area, in square feet

$$(1.7TL_{WL}) + (\frac{35\,D}{T}) \qquad\qquad \text{Eq.3.138}$$

T=Vessel draft, in feet, $L_{WL}$ =Waterline length of vessel, in feet, D=Displacement of ship, in long tons, $C_{xca}$ = Longitudinal skin-friction coefficient

$$\frac{0.075}{(\log R_n - 2)^2} \qquad\qquad \text{Eq.3.139}$$

$R_n$=Reynolds number = $V_c L_{WL}\cos\theta_c/\gamma$, $\gamma$ = Kinematic viscosity of water ($1.4 \times 10^{-5}$ square feet per second), $\theta_c$ = Current angle.

Current yaw moment is determined using the following equation:

$$M_{xyc} = F_{yc}(\frac{e_c}{L_{WL}})LWL \qquad\qquad \text{Eq.3.140}$$

Where: $M_{xyc}$ = Current yaw moment, in foot-pounds, $F_{yc}$ = Lateral current load, in pounds, $(\frac{e_c}{L_{WL}})$ = Ratio of eccentricity of lateral current load measured along the longitudinal axis of the vessel from amidships to vessel waterline length
    $e_c$=Eccentricity of $F_{yc}$, $L_{WL}$ = Vessel waterline length, in feet. The value of $(\frac{e_c}{L_{WL}})$ is given as a function of current angle, $\theta_c$, and vessel type.

## 3.13.10 Traffic Separation Scheme (TSS)

TSS can be modeled from deduced based on standard deviation equations below (Figure 4.46).

$$f_{south}(x) = \frac{1}{\mu\sqrt{2\pi}}e^{-\frac{1}{2}}(x-\frac{12}{\mu}) \qquad\qquad \text{Eq. 3.141}$$

$$f_{north}(x) = \frac{1}{\mu\sqrt{2\pi}} e^{-\frac{1}{2}}(x - \frac{12}{\mu})$$

Eq. 3.142

Due to symmetry one side analysis is given by:

$$F_1 = \int_{12}^{25} \frac{1}{\mu\sqrt{2\pi}} e^{-\frac{1}{2}}(x - \frac{12}{\mu})(x).\,dx = \int_{12}^{25} fsouth(x).\,dx$$

Eq. 3.143

Figure 3.39 shows the TSS scheme model diagram for inbound and outbound vessels.

Violating traffic $N_1 = NmF_1$ ships per sec.

Eq. 3.144

The traffic density $\rho = \frac{N_i}{W}$ ships per second$/m^2$

Eq. 3.145

Probability of encounters $P_i = 2B.D._{ps}$ 1/ passage

Eq. 3.146

For non violating traffic $F_1 = \int_{12}^{25} f_{North}(X)dx$

Eq. 3.147

Violating traffic $Nm = N_m.F_2$ Ships per year

Eq. 3.148

For violating traffic

$$F_2 = \int_{12}^{25} f_{South}(X)dx$$

Eq. 3.149

Expected number of head on collision:

$Fa = P_i.P_c.N_2$      Collision per year

Eq. 3.150

Due to symmetry, collision analysis should be considered twice.

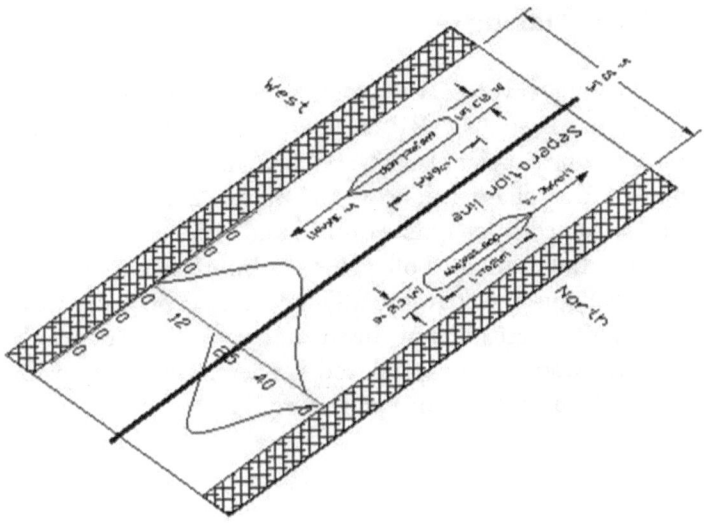

a. Separation system

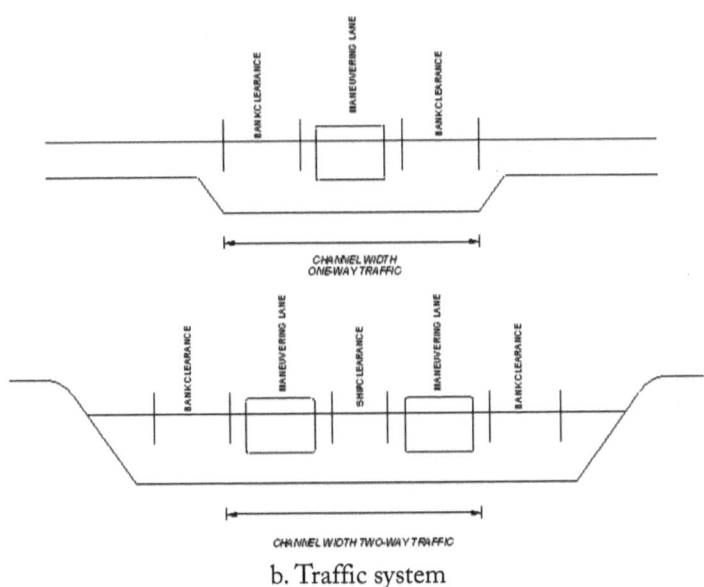

b. Traffic system

**Figure 3.46**: Traffic separation scheme

## 3.13.11 Decision Making

Decision making involves discussion of hazard and associated risks, review of RCO that keep ALARP curve in acceptable region, compare and rank RCO based on associated cost and benefit. Data collected organized and presented in this chapter. The result of the model implementation is presented and discussed in chapter Four. Sound risk based model for reliable IWT could form the strategy policy for transport sector in implementing inland water development that broadly address, economic, infrastructure, safety, environment, management institutional, legal framework and the use of intermodal system. The proposed model could be transformed into software that will be useful for universal usage.

# 4

## Result

### 4.1 Safety Risk and Reliability Model

The model is use to determine the probability of failure or occurrence, risk ranking, damage estimation, high risk to life safety, cost benefit analysis, sustainability and acceptability criteria of IWT. Seasonal trends are model from probabilistic result. Accidents that are considered intentional are negligible in the model. Environmental condition and traffic volume fluctuation is considered negligible in the model. Table 4.1 present the generic waterway system risk are deduced from the system presented in Figure 3.9 and Figure 4.1.

### 4.2 Result of Waterway System Functionality and Standard Analysis

Figure 4.1 shows ship movement in waterways. Figure 4.2 show block diagram of simplifies system with the input (Po, Td, S, E) and output (Rc) and feedback that represent sensing and detection point. Figure 4.2 represent the deduce system block diagram from the schematic. Table 4.1 shows deduced result from the physical system assessment.

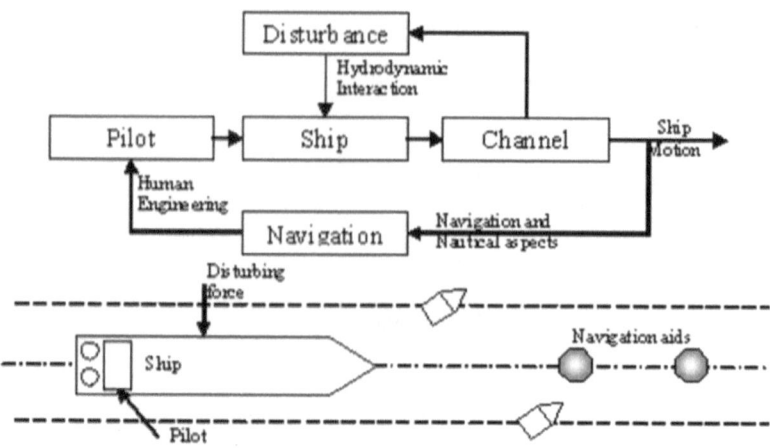

**Figure 4.1:** Inland waterway system

**Table 4.1:** System definition and risk assessment

| Ship | Channel | Traffic | Disturbance | Human/ organzation | Consequences | |
|------|---------|---------|-------------|--------------------|--------------|--|
| | | | | | Short term | Long term |
| Deep draft | Visibility | Traffic density | Wind | Watch keeping | Volume of cargo | Damage to properties |
| Shallow draft | Passing vessel | Deep draft vessel | Visibility | Human Interaction | Volume of petroleum | Damage to economic |
| Propulsion | Channel bottom | Shallow draft vessel | Current | New technology | **Volume of chemical** | **Damage to environment** |
| Navigation | Waterway complexity | Leisure Vessel | Wave | Traning | **Volume of cargo** | **Damage to Health** |
| Crew | Obstacle | Fishing vessel | Tide | Attitude | **Volume of passenger** | **Injury** |
| Flag | Regulation | Leisure Vessel | Mooring | Manual | **Cargo operation** | **Death** |

## 4.3 Result of HAZID (Collision Risk Scenario Qualitative Assessment)

**Table 4.2:** HAZOP assessment

| No | Guideword | Description | Causes | Safety measure | Risk |
|----|-----------|-------------|--------|----------------|------|
| 1 | No Pitch | No rotational energy is transformed | operation, control mechanism, alignment failure | address by 2, 3, 4, 5 | 3 |
| 2 | No blade | No rotational energy is transformed | Object in the water break the blade | implementation of propeller protection such as grating jet, sail in ice free water, +7& 9 | 4 |
| 3 | No control bar | All blade on random pitch, loss of operational control | material weakness | improve design and construction | |
| 4 | No crank wheel | On all blade have independent pitch | material weakness | improve design and construction | 3 |
| 5 | NOT enough material strength | part of propeller breakdown | wrong design, corrosion or cavitations, alignment different pitch, extra load on bearing | validate propeller design , catholic protection, appropriate propeller material , test the propeller against cavitations ,periodic alignment adjustment | 4 |
| 6 | MORE pitch than optimal | Too heavy load on propulsion system. Cavitations | operation failure | surveillance, increase operator competency | |
| 7 | LESS pitch than optimal | Too little load on propulsion system. Cavitations | operation failure | surveillance, increase operator competency | 2 |
| 8 | LESS draft than allowed | Propeller I not sufficiently submerged. Loss of Thrust | operation failure | surveillance, increase operator competency | |
| 9 | LESS depth than necessary | Propeller hit the ground and it is damaged | operation failure | technical equipment, surveillance, increase operator competency | 3 |

Figure 4.3 shows components of PrHA. Risk contributing factor to collision are presented in Table 4.3.

Table 4.3: PrHA for collision event

| Hazards/ Events | Causes | Probabilities analysis | Consequence analysis | Risk | Possible measure to reduce risk |
|---|---|---|---|---|---|
| Powered vessel to Structure collision | Visual restriction at night | Probable (4) | Death/serious injuries (D) | 4D | Night navigational aids made available |
| Moored vessel collision to structure | Unaware | Remote (2) | Death/ disability (D) | 2D | Installation of proper signage. Site identified with buoys. |
| Passing vessel to vessel collision | Too close/ Net caught under dredger | Probable (4) | Probably major injury (C) | 4C | Restrict navigation at nearby site. Install buoys. |
| Propulsion failure | Break down | Probable (4) | Probably major injury (C) | 4C | Regular maintenance |
| Mooring failure | Weather damage | Probable (4) | Probably major injury (C) | 4C | Make assessment on weather conditions. |

# 4.4 Result of Accident Frequency Analysis (AFA)

Figure 4.2 to 4.5 describe the result deduced accident frequency analysis.

## 4.4.1 Impact Probability and Angle of Impact at Changing Velocity of Ship

Figure 4.2 a and b show accident impact probability correlation with speed and angle of impact. The generic model developed is flexible and it can be applied to any case. The model is tested on Langat by simply including the variables that match Langat data to deduce the behaviour in the model. Figure 4.2 to 4.5 represent intermediary result of impact probability and of impact and collision situation that reexamine need to

use acceptable angle of collision, generic collision situation and impact
probability window. Also in the analysis that is presented in this chapter,
combine are intermediary results for further, the combine plot apply to both
case study and generic. Uncertainty analysis is also intermediary result for
the model. Qualitative analysis is also, another form of intermediary result
in SARM model, they are used to reexamine causal factors, and their result
depends on channel to channel.

The result from intermediary analysis applies to both generic
model and case study validation result. Figure 4.2 a and b show impact
probability and velocity relation, Figure 4.2a emphasize on respective
accident collision situation analysis. Head-on, overtaking and circular
collision is likely to happen in the same direction while specific angle is
likely to occur in any or different direction. It is also observed that impact
probability is likely to be high and frequent at speed less than 3 and higher
than 7. Figure 4.2b shows that impact probability depends on the speed
of ship. Figure 4.2b shows that impact become more intense at 50 degree,
60 degree and when the speed is higher than 12 knot. The risk become
extremely increase at 12 knot. The graph also shows that the risk of
accident of occurrence is possible to rise when the ship is moving at much
lower speed. These numbers and derived regression equation provided
in Table 4.8 are useful for limit definition for the channel. The plot in
Figure 4.2 shows the operating working speed for random collision, where
moderate speed required is observed below 12knot. Average speed below
10 knot gives minimum impact, a ground that match with maximum speed
limit in waterways.

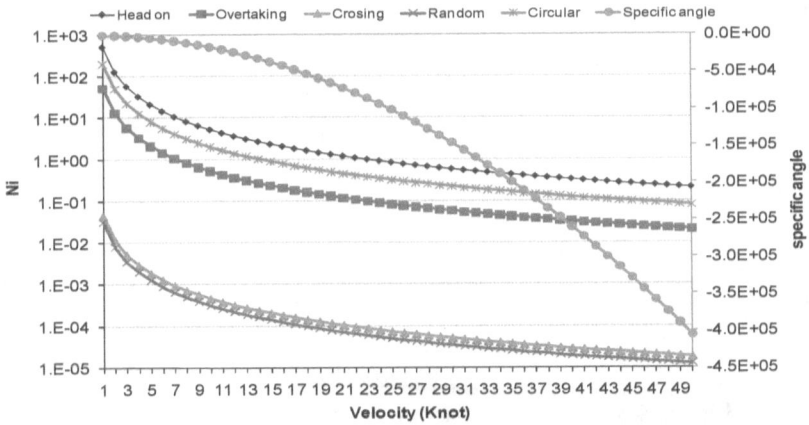

a.

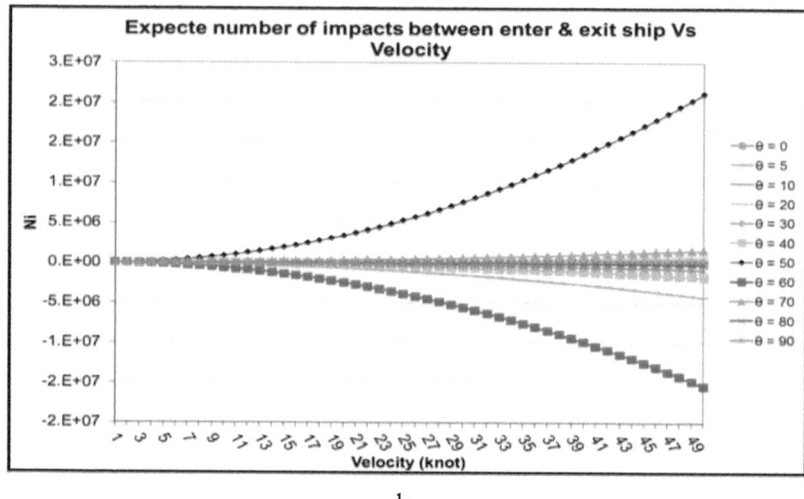

b.

**Figure 4.2:** Impact probability and velocity

Figure 4.3 shows that impact probability and impact angle plot. Figure 4.3 shows accident impact probability correlation with speed and angle of impact. Highest impact is observed at Fa of 5E+10 and 22 degree. Figure 4.3 shows that impact probability is likely to be high at collision angle between 45°, 60° and 90°. It is observed that collision impact probability is intense at 20° to 30° and favor high impact probability.

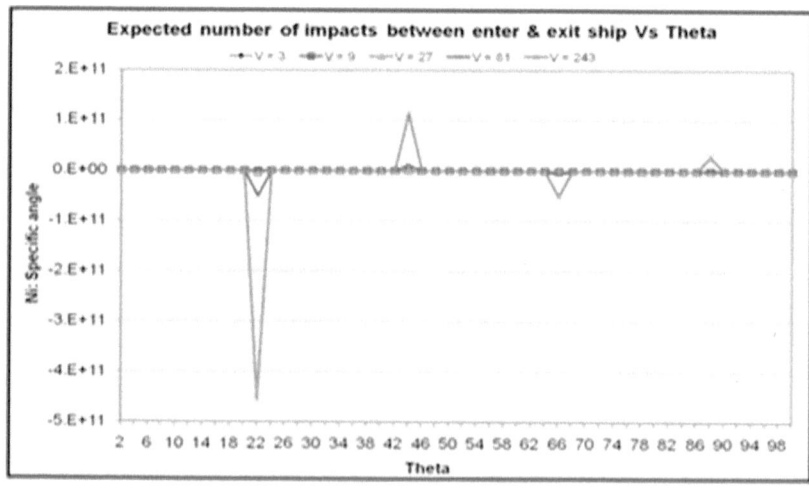

**Figure 4.3:** Accident impact probability and angle of impact

Figure 4.4 shows Ni and velocity for random collision situation at changing number of ships. Figure 4.4 shows impact probability relation with different number of ships. The plot show that the higher the number of ship the higher the expected impact. The highest impact is expected at impact acceptability, while breakdown is expected at speed below 3 and speed above 7. In Figure 4.4 it is observed that, as the number of ship in channel increase, the impact probability in the channel increase. Figure 4.4 also shows that, there is higher risk and potential rise in Ni as the number of ship in the channel increase, 1 to 2 number of accident impact probability is likely to occur at 12 number of ships for 10-12 knots. For ship in the channel navigating at 4 to 7 knots, there is likelihood of 0-1 impact probability for acceptable frequency of accident (1e-3).

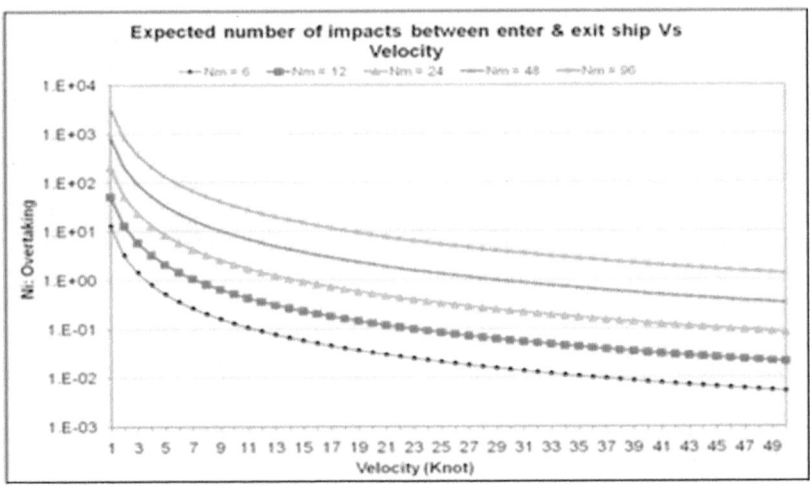

**Figure 4.4:** Impact probabilities Ni Vs velocity and Nm

## 4.4.2 Result of Collision Accident Impact Probability at Different Collision Situation

Figure 4.5a shows that head on collision has the highest risk potential. Additional analysis of impact probability is shown in Figure 4.5a and b, Figure 4.5a and b show comparison of collision situation, Figure 4.5a shows expected number of impact, accident energy at variable speed plot for head on collision increases, as the number of ships in the waterway increases. Figure 4.5a shows that head—on collision prove to be more risky. Figure 4.5a also shows that, for overtaking collision, is more likely to pose high consequence to the channel. Figure 4.5b also shows that head on collision has the highest risk potential as the curve touch high risk area. Derived regression equation could be used for further future test. Figure 4.5b also revealed that, overtaking collision situation has less risk than other collision situation. Figure 4.5b shows that the graph of accident impact probability increase with increase number of ship and also increases in speed; optimum speed with low impact probability is 4 knot. Six numbers of ships navigating at speed of 4-5 knot will have less or no impact probability. The trend shows that high impact situation takes many years to happen.

Frequency analysis is estimated from traffic density, expected number of collision per passage (Ni), necessary period for ship to pass the fairway (T), primary data for the generic model is extracted from PIANC criteria for waterway classification (Table 3.18). The result of frequency probability (Fa) analysis observed is checked and compared with frequency acceptability criteria index. Maritime industry risk index acceptability is use to measure risk. In SARM model, Low risk is considered at Fa values of 10e-4 or 10e-6 and Fa high risk is considered at 10e-2 to 10e-3, and acceptable risk is 10e-4. Above exponential 3 is consider risky, while below exponential 6 is considered negligible. Risk for different collision situation is examined, at speed of 23knot there is low risk for random and crossing (Fa=10e-6). At the same speed there is high risk for overtaking, circular and head on collision (Fa=10e-2).

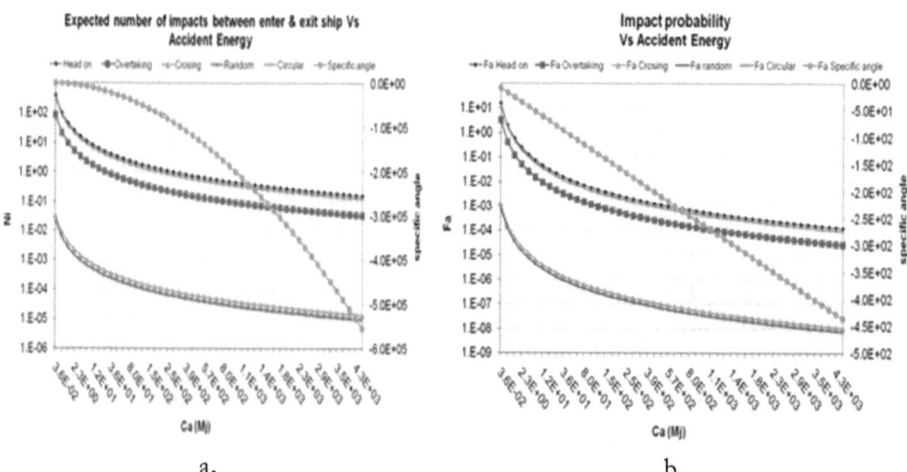

a.                                    b.

**Figure 4.5:** a. Collision risk potential for different collision situation, b. Impact probability Vs accident probability.

Result of model for accident frequency analysis is as followed:

### 4.4.3 Accident Frequency and Speed

Figure 4.6 shows the plot of frequency and speed when the number of vessel is changing. Figure 4.6 shows the plot of frequency and speed when the number of vessel is changing. Figure 4.6 shows accident frequency and velocity relation. Unacceptable speed of ship is observed at 10-12 knot for 12 number of vessel in the channel. Reduce speed less than 3-4 knots are considered risky for 12-48 numbers of ships, at that moment the Fa is 10e-3. It is also observed that speed of 7 knot is acceptable for 12 number of vessel, at accident frequency acceptability of 10e-4.

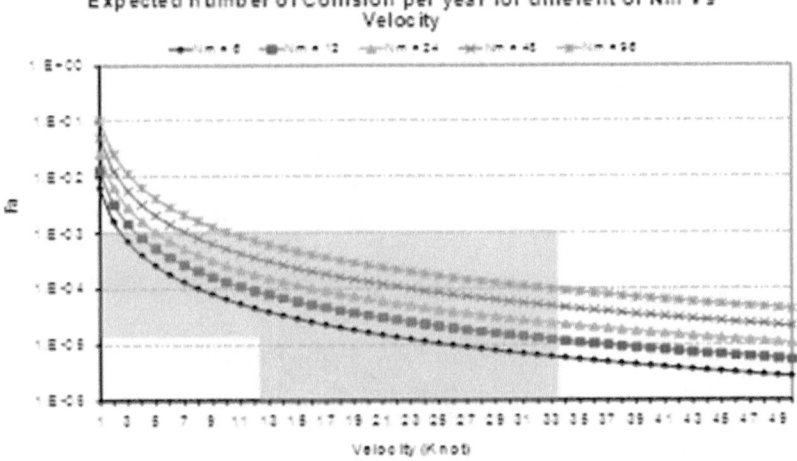

**Figure 4.6:** Accident Frequency and Speed

### 4.4.4 Accident Frequency and Different Number of Ship

Figure 4.7 shows accident frequency and velocity relation. Figure 4.7a shows that at changing traffic of 12 number of ship at speed of 9 are acceptable in the waterway. At maximum acceptable speed of 12 knot, maximum of all number of ships except six ships can steam through the channel. At high risk level, 96 ship should sail at speed of 9-10 knot to be safe and 48 ships at 7 knot. At 10e-4, 48 ship and 12 ships is lower. At lower number of ships, ships can sail at lower speed to be safe. Figure 4.7b shows plotting for the waterway accident frequency and number of ships, at changing angle of course. Figure 4.7b shows that, increase in number of ship pose high risk at all angles, angle between 50-70 posed much higher risk, at all angle acceptable number of ships leading less frequent accident is between 7 to 10 ships.

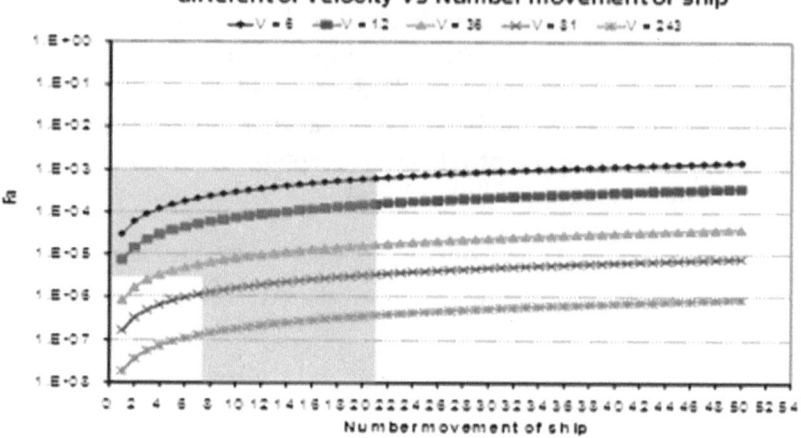

**Figure 4.7:** Accident frequency and different number of ship and speed

## 4.4.5 Accident Frequency Vs Width

Figure 4.8 show accident frequency against changing width and beam of the channel. Figure 4.8a and b shows the plot of frequency and width, both plot are the same plot that show accident frequency against and width of the channel at changing speed. The graph shows where the maximum width the channel and the corresponding speed become risky. From the plot, it is observed that as width of the channel decrease, there is higher risk. The maximum width considered for the biggest channel by PIANC is 102m, and speed of 12 knot is in safe region in this model. At accident unacceptable frequency of 10E-2, acceptable width is 774m for 12knot speed, here the maximum width for 12knot is 270m but Fa is 10e-3 for 24 knot speed with maximum width of 774m. The Figure also revealed that navigation at less than 3 knot at width up to 98m pose high frequency of accident. At speed of 4 knot wide of 55 and beam of 35 is at Fa of 3e-2 with 96 ship, this is consider risky, as accident is expected to be more frequent at that point.

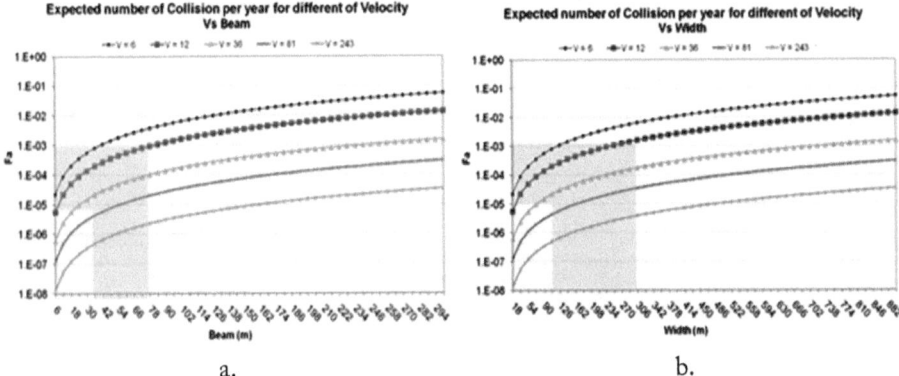

a.

b.

**Figure 4.8a:** Accident frequencies and

**Figure 4.8b:** Accident frequency and beam of channel width of vessel

## 4.4.6 Result of Accident Frequency, Velocity of Ship and Angle

Figure 4.9 shows accident frequency, changing Nm and angle. Beam plays a very important role in decision support for waterway risk. Beam of ship represent 3 times the width of the channel. Figure 4.9 shows accident frequency and changing beam of ship, number of ships and angle. Figure 4.9a shows how much number of ships is allowed in the channel at respective beam to maintain channel integrity. At moment of unacceptability accident frequency of 10E-2, maximum Nm of 96, beam of 115m is clinched. At acceptability of 10e-4, beam of 34m is suitable for 24 ships, while beam of 18m is suitable for 96 ships, less beam is prefer for more ships. 96 ships at frequency unacceptability of 10e-2 will require beam of 55m, at same time 6 ships with beam of 50m is allowed to sail the channel. At frequency acceptability level of 10e-3, beam of 55m is allowed for 6 numbers of ships, for 96 ships the same beam is at high risk region 10E-2, allowable beam for such number of vessel is 15-20m in the waterway. Additional analysis of waterway beam is presented in Figure 4.9b, it shows Fa, beam and angle interaction, collision frequency is the same at all beam, but acceptable beam for all angles is 30°, beyond 30 °, it is to have high risk. At frequency acceptability of 10E-4, beam of 10-40m is acceptable at all angle of collision. PIANC recommended 34m as the largest bean for ship of class VII waterways classification, at this value, the ship can navigate in all angle. Number of ships and speed can be recommended for difference beam of ship.

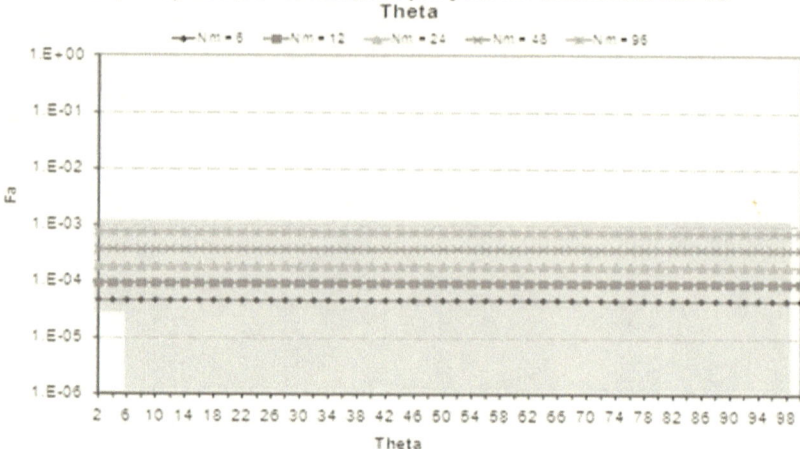

**Figure 4.9:** Accident frequency and angle

## 4.6.7 Result of Combined Graph of Waterways Variables

Figure 4.10 Shows combined graph of accident variables. Figure 4.10 a, b and c shows combined graph of accident variables. Figure 4.10a, b, and c shows similar trend. For accident at 10e-3, the graph shows that small beam and small number of ships pose low risk, whereas too low speed poses high risk. At unacceptable frequency of 10e-2, there is agreement of speed of 12knot and 12 ships number of ships and 78m beam, and also 6 number of ships 55m beam and beam of ships ship agreed.

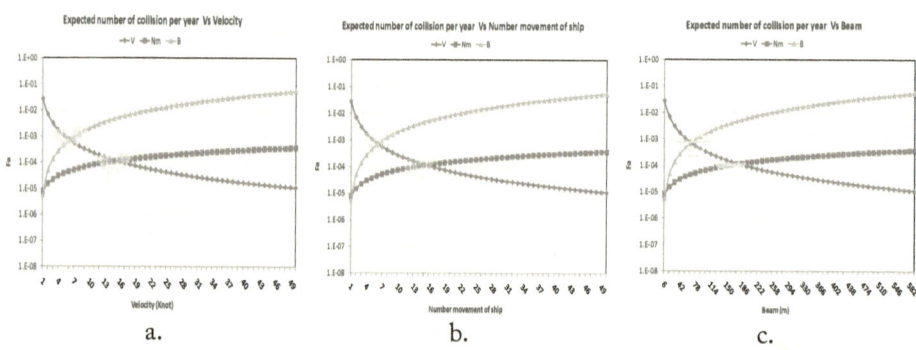

**Figure 4.10:** Combined graphs for frequency analysis with changing channel variables, a: velocity, b: number of ships, c: beam

## 4. 5 Result of Accident Consequence Analysis

Collision risk is a product of the probability of event occurrence and consequence of the event. Occurrence of accident results to different kind of losses like damage to infrastructure, loss of life, injury, third party and economic losses. The deduced consequences to IWTS are dynamic. The acceptability criteria is transformed and also could be represented in form of energy, oil spill, and cost of fatality. This study modeled energy consequence; the deduced energy value is translated to oils spill or carbon dioxide released (Refer to Table 3.27).

### 4.5.1 Power Collision Velocity and Changing Mass

Figure 4.11 a, b, and c show the energy consequence velocity plot for different mass of vessel, volume of collapse and length of collapse relation for power collision. Figure 4.11 to 4.13 shows the energy consequence and velocity plot for different mass of vessel in waterway. In Figure 4.11, mass, volume of collapse and length of collapse and energy against velocity for powered collision is modeled. In Figure 4.11a, at speed of 12 knot energy release for all 300MJ is obtained for mass of 30,000t, in Figure 4.11b, 60, volume of collapse is observed for the same velocity, while in Figure 4.11c, length of collapse up to 7-10m is observed for the same speed. At target speed of 4 knot, Lp is about 0.4m, Vc is about and E1 is about 50MJ, all mass of vessel can be accommodated within acceptable region. At mass of vessel is 18,000t. 1m length of collapse can occur at 7 knot with 20 cubic meter volume of collapse and 400 MJ release of energy.

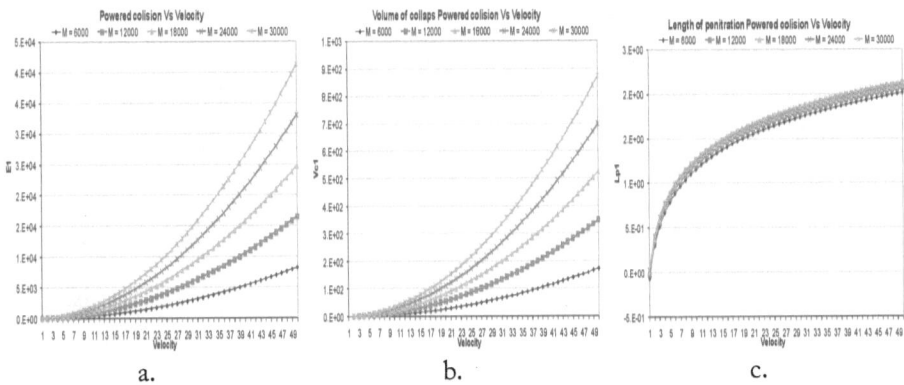

**Figure 4.11:** Energy consequence velocity plot for different mass of vessel, a: energy, b: volume of collapse, c: length of collapse

## 4.5.2 Drifting Collision Energy, Velocity and Mass

In Figure 4.12 shows drifting collision energy and velocity at changing mass. Figure 4.12 represent drifting collision result, similar trend is observed except that, small vessel are more likely to have much higher impact compare to large vessel. It is observed that more length of collapse can occur, but less energy and volume of collapse.

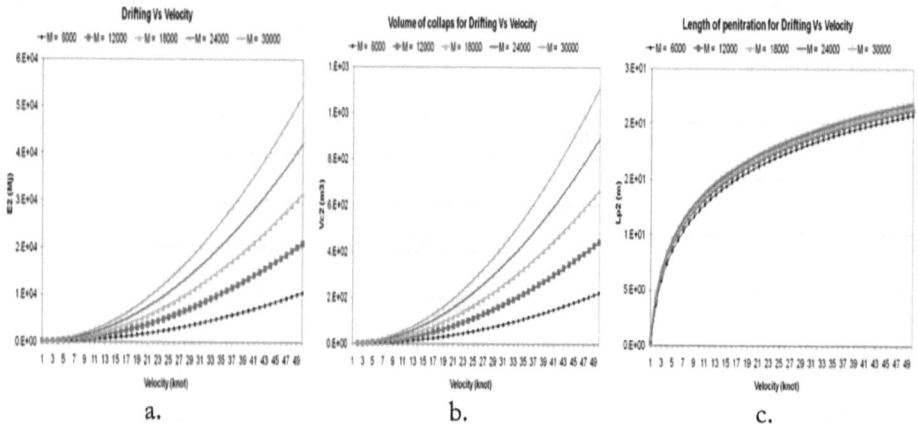

a.                               b.                               c.

**Figure 4.12:** Drifting Collision Energy and Velocity at Changing Mass:
a: energy, b: volume of collapse, c: length of collapse

## 4.5.3 Vessel to Vessel Collision Velocity and Mass

Figure 4.13 shows the vessel to vessel collision. For vessel to vessel collision in Figure 4.13, similar trend is observed thus, much high release of collision energy, volume of collapse and length of collapse is observed. 7 knot speed could result to 5m length of collapse, 100 volume of collapse and 1000MJ released of energy.

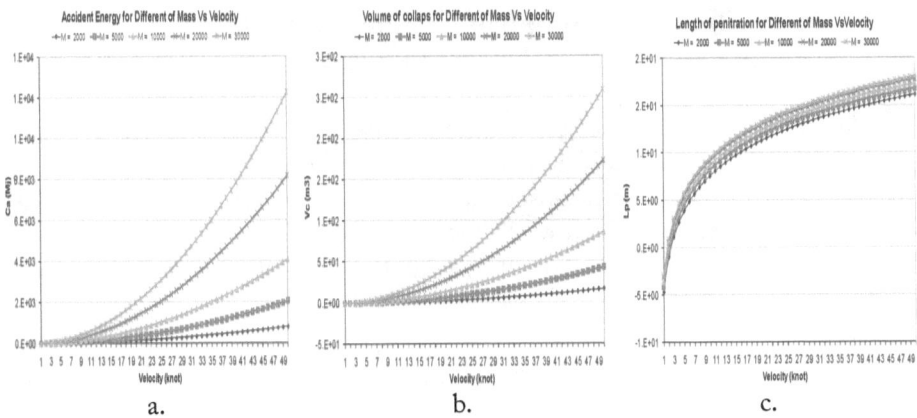

**Figure 4.13:** Vessel to vessel collision energy and velocity at changing mass
a: Energy, b: Volume of collapse, c: Length of collapse

## 4.5.4 Vessel Collision Velocity at Angle of Collision

Figure 4.14 show that collision is much harsh at angle 5 degree. Figure 4.14 shows additional collision volume of collapse angle of collision and velocity relation for vessel to vessel. In Figure 4.14a it is observed that collision is much harsh at angle 5 ° to 10 ° and 20 ° to 3 ° 0 degree. In Figure 4.14b it is also observed that at angle of 5 °, more length of collapse is expected, but less volume of collapse.

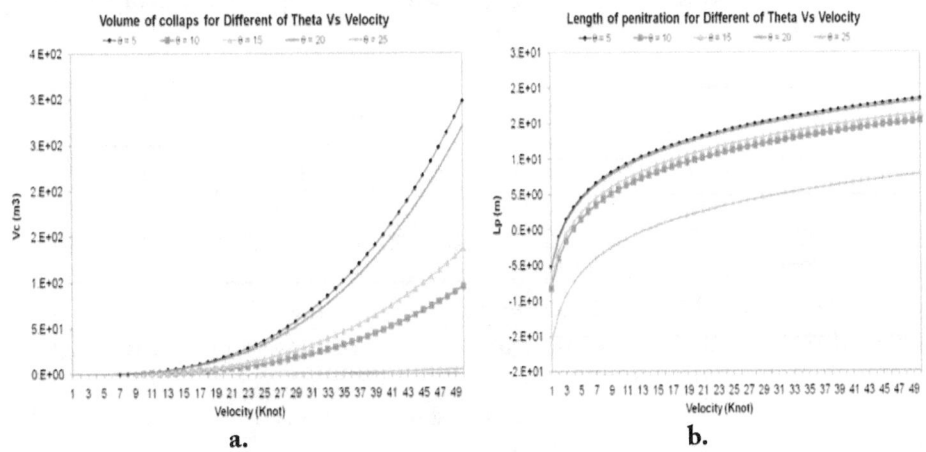

**Figure 4.14:** Vessel to vessel collision energy and angle of collision

185

## 4.5.5 Energy Release, Volume of Collapse, Length of Collapse and Mass of Powered Collision

Figure 4.15 shows power collision mass energy relationship. Figure 4.15a, b, c shows power collision situation with mass, energy of collision and velocity of interaction, it is observed that at maximum released of energy 400MJ, and maximum speed of 12knot, at acceptable mass of 10,000t, the volume of collapse also correspond to 60and the length of collapse of about 100m.

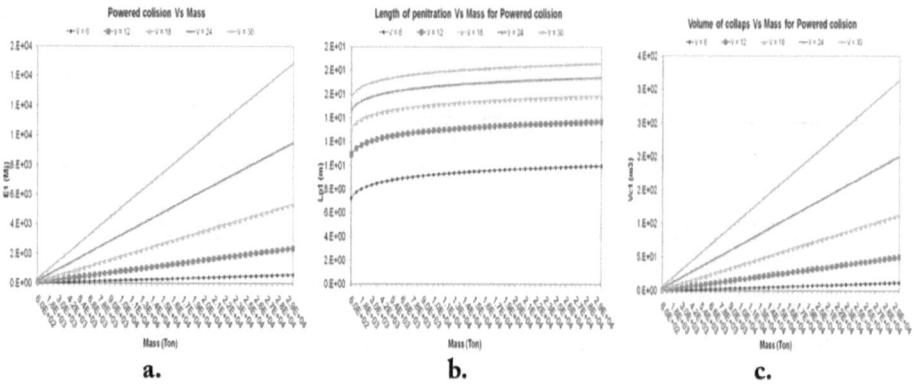

a.       b.       c.

**Figure 4.15:** Powered collision energy and velocity at changing mass

## 4.5.6 Energy Release, Volume of Collapse, length of Collapse and Mass of Drifting Collision

Figure 4.16 show drifting collision energy velocity and mass. From Figure 4.16a, at maximum mass of 27 to 30t, acceptable maximum speed is 6 to 12knot, at the same point in Figure 4.16b, the volume of collapse is less that 50and the length of collapse is less than 7m. The trend generally revealed that higher speed leads to more volume of collapse and length of collapse. Figure 4.16 show situations with drifting collision situation, similar trend is observed, like the powered collision, but there is slight increased in released energy but less volume and length of collapse compare to powered collision. At 2000MJ, 30,000t is at 12 knot could lead to 20m length of collapse and 500 of Vc. There are differences in release energy between the three situations as mentioned above.

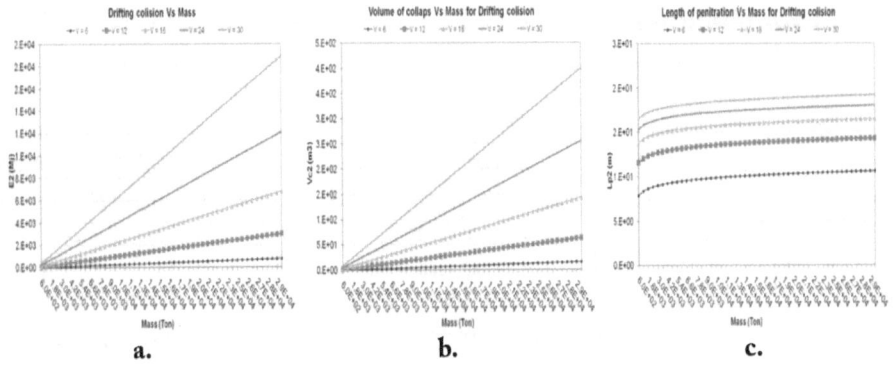

**Figure 4.16:** Drifting collision energy velocity and mass

Figure 4.16 and 4.17 further show similar trend for drifting and vessel to vessel collision as discussed above in Figure 4.15 for powered collision.

## 4.5.7 Energy Release, Vc, Lp for and Mass of Ship to Ship Collision

Figure 4.17 shows situations with vessel to vessel collision. Figure 4.17 shows situations with vessel to vessel collision; similar trend is observed, with slight decreased in released energy at highest speed, but less volume and length of collapse compare to powered collision. 30000t vessel could lead to 7m length of collapse and 20e+12J released of energy (4.4x10e+11J). In the analysis for vessel to vessel collision, catastrophic volume of collapse can occur at 90 cubic meter with of vessel mass up to 12,300t. Also it is observed that vessel up to 19,000t could leads catastrophic release of energy up to 19,205MJ. Acceptable damage penetration of 7m start to occur at 60MJ, and damage penetration of 10m occur at 400MJ consequence energy.

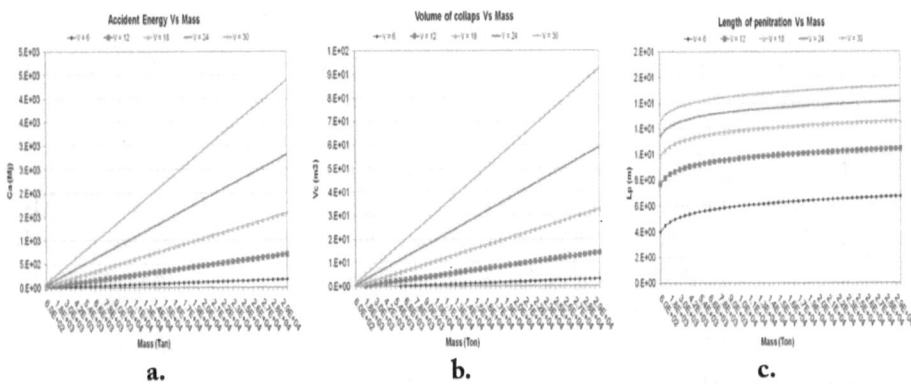

**Figure 4.17:** Vessel to vessel collision energy and velocity at changing mass

## 4.5.8 Combined plot of Consequence, Volume of Collapse and Length of Collapse

Figure 4.18 a and b shows variation between mass and accident energy, Lp , Vc. Figure 4.18a, b, c, shows situation of powered collision, where length of collapse starts to become critical at 10m, 60 cu. From the graph it is observed that at any number of ship, the system damage follow 100 percent trend for Vc, Lp. Except that the level of damage differ. Figure 4.18a shows that the at maximum energy acceptability of 400MJ, 7-10m length of collapse could occur at the same instance, in Figure 4.18b, length of collapse of 15m could is observed, while in Figure 4.16m 15m to 16m is observed. The earlier comes with low release of energy but high volume of collapse while the later is opposite.

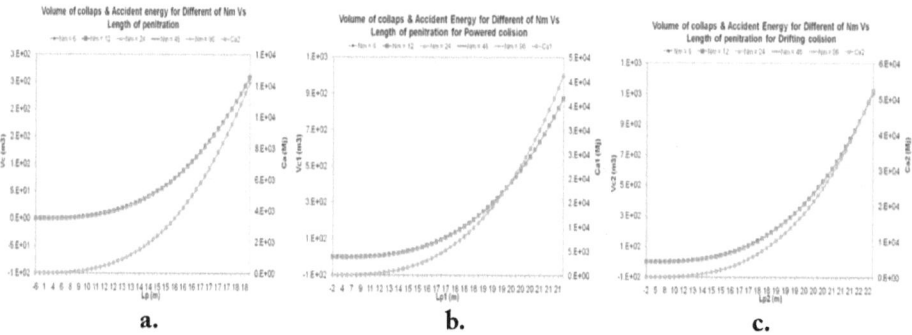

**Figure 4.18:** Combine plot of energy, volume of collapse and length of collapse

## 4.6 Result of ALARP

The following shows result deduced from ALARP analysis. The consequence analysis is followed by ALARP risk influence diagram that is based on ALARP principle, here the two main component (frequency x consequence) of risk is analysed simultaneously and plotted against each other to measure overall risk in IWTS. Risk acceptability criteria presented in Table 3.22 and Table 3.24 are used to measure the risk level for the ALARP analysis. Table 3.23 shows risk acceptability criteria traditional rating. In the modeling of the potential losses, classification of performance indicators between those phenomena that are ever present during IWTS system lifecycle called systemic (probability = 1) and those which occur rarely in the ship lifecycle called random (probability < 1).

Risk acceptability criteria are established in many industries and regulations to limit the risk in the system (Table 3.24). Consequence thresholds, with priority of value choice are awarded. The highest consequence tripped in order of priority give the overall consequence with descriptor of catastrophic, major and insignificant for 1 for people, 2 for Infrastructures, and 3 for values (Figure 3.24). Risk is never acceptable, but the activity implying the risk may be acceptable due to benefits to safety, fatality, injury, individual and societal risk, environment and economy. Societal criteria risk (Table 3.22) is used by increasing number of regulators.

A further classification of potential losses can be defined in terms of the consequences impacting the ship or other assets, humans (crew) on board, and the environment. The systemic events occur with probability of one, the consequences that follows are still uncertain. The systemic events as well as the random events are treated in a risk based design, it summarizes the elements which are also dynamic economics, society, and environments are taken into consideration.

The risks associated with these elements are both systemic phenomenon (p = 1), or random phenomenon (p < 1). Design has conventionally been concerned with systemic economic risks through first and operating design costs, while owner's hedge their accidental risks via insurance. After decision based on ALARP consideration other channel complexity and uncertainty, reliability and sustainability analysis are considered for adjustments on waterways requirements. Regulation is concerned with the risks to life, the property of third parties and environmental protection. Only in recent years has the systemic risk to the environment become a significant concern, with ever increasing importance. The following section presents the ALARP result of the SARM model.

### 4.6.1 ALARP Graph for Accident Frequency and Consequence Energy

The ALARP graph for accident Frequency Vs Consequence is based on ALARP principle that is adopted for maritime industry risk acceptability. Figure 4.19 shows ALARP of combined accident variables. Figure 4.19 shows the ALARP graph for accident frequency and consequence. The risk is measured based on ALARP principle that is adopted from UK HSE offshore QRA for maritime industry FSA risk acceptability criteria. The ALARP acceptability diagram tolerate optimum accident acceptable probability of 1e-03, optimum amount of energy has optimum acceptable energy of 100e+6 to 150E+6. The curve for generic model is at the lower portion of ALARP which means that risk result lies in safe zone of ALARP graph acceptability as compare with Figure 3.25a (Risk acceptability criteria). The graph show combined curve for risk related to waterway variables, the waterway parameter fall within acceptable region of the ALARP graph given in Figure 3.25a, accident frequency is between 11e-5 and 1e-3 while Ca is between 1MJ to 33MJ. V and Nm agree at lower risk region, while V and B agree at much higher risk region.

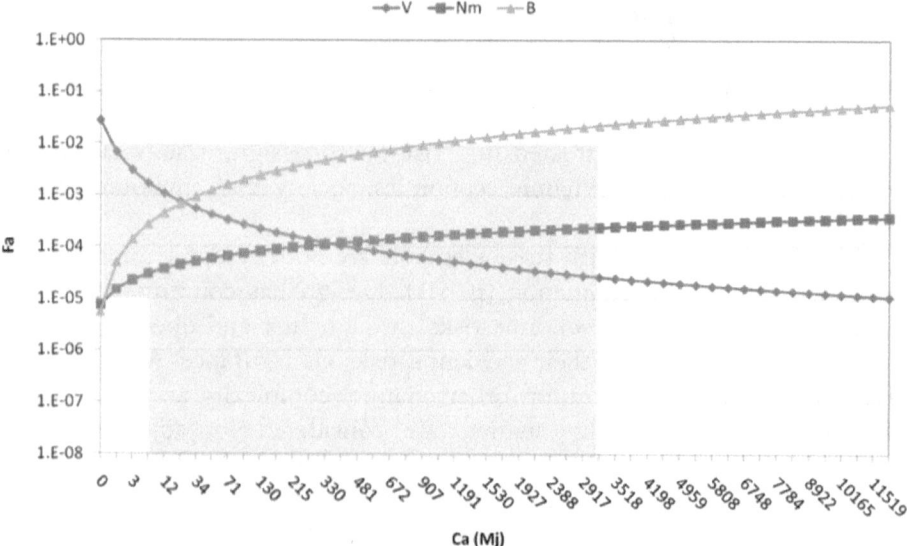

**Figure 4.19:** Accident frequency Vs Consequence

## 4.6.2 Accident Frequency Vs Consequence Energy and Number of Ships

Figure 4.20 show the risk estimation for changing number of ships in the channel. Figure 4.20a and b show the risk estimation for changing number of ships in the channel; it is observed that at accident frequency acceptability of 10e-4 years 100 to 150MJ of energy is released with 96 number of ship in the channel at speed of 12knot. 96 number of ship in the channel fall between frequency and consequence acceptability. Risk situation for accident frequency at changing number of ships indicate that risk become higher with more number of ships. Figure 4.20a shows the risk estimation for changing number of ships, at 1e-3 acceptable accident rate in the channel; it is observed that, at maximum accident frequency of 10e-2, 100MJ of energy could easily be released with 96 number of ship in the channel. Figure 4.20b shows cross-plotting graph for the same Nm. 9 knot is good for 12 to 15 ships at Fa acceptability of 1e-3, with roughly 30MJ Ca.

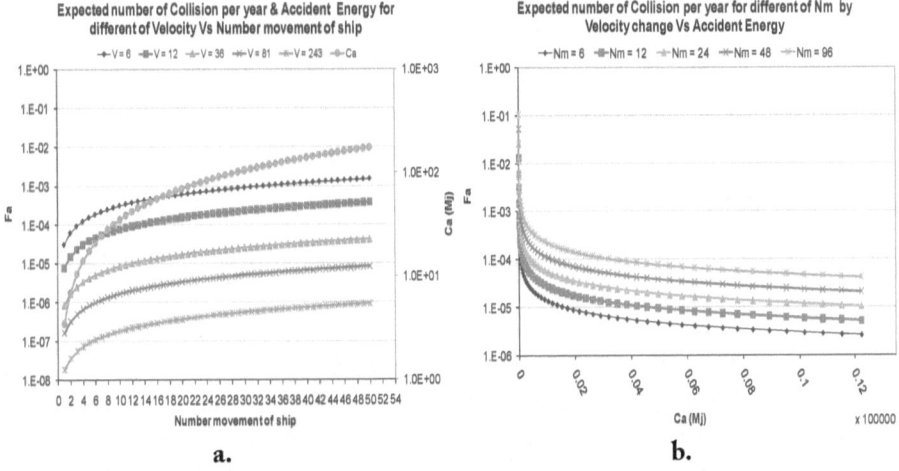

**Figure 4.20:** Accident frequency, consequence and changing Nm

## 4.6.3  Accident Frequency Vs Consequence Energy and Speed

Figure 4.21 shows ALARP of speed. Figure 4.21and b show accident frequency and consequence with speed and number of ship interaction. Figure 4.21 shows that ships, from 1to100 can operate on the channel, but that comes with higher consequence within acceptable risk limit of (1e-3 and 400MJ). And the consequence increase with increase in number ships. Figure 4.21a shows cross plotting shows that at risk acceptability accident frequency of 1e-4, speed of 12 knot is good to maintain 96 ships, but the consequence as high as 1000MJ. At acceptable 150MJ release of energy, 48 ships are good for 12 knots. It is also observed that lower speed favored frequent accident occurrence frequency and higher consequence. Also lowest energy release of 50MJ can be maintained for 12 ships at about 11to12 knot. For the release of minimum about 50MJ, 96 number of ship falls within acceptable region of ALARP. Figure 4.21b shows cross plotting analysis of risk agreement of 12 to 24 ships at 10 to12 knots is at 1e-4—1e-3. For generic model, speed of 2-3 could be too dangerous. Speed of 4 is acceptable. 96 ships can be accommodated at that speed, but, at high risk acceptability of 1e-3.

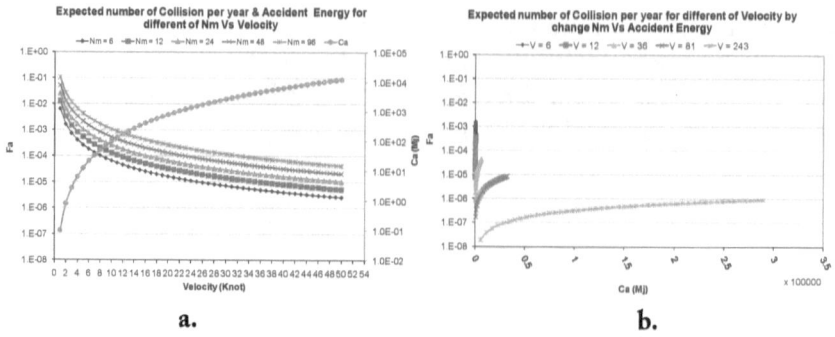

a.                 b.

**Figure 4.21:** Accident frequency, consequence and speed

## 4.6.4 Accident Frequency Vs Consequence Energy and Beam and Width

Figure 4.22 shows that risk become higher at large beam of ship. Figure 4.23 shows cross plotting for the effect of width and beam of ship. Figure 4.22 and Figure 4.23 show risk of beam and width interaction. It is observed that the risk becomes higher at large beam of ship. Figure 4.22a shows ALARP curve for beam changing. Figure 4.22b shows further analysis of the curve through 2 dimensional plotting of Fa and Ca. Largest beam of ship is maintained recommended by PIANC largest beam 35m. Increase in beam of ship could lead rise in risk. At accident acceptability for beam of (35m) is used for the model development, speed up 12knot, 96 ships can be maintained but at high cost up to 1000 MJ released of energy. For acceptable release of energy between 150 to 400 MJ at 35m beam, 24 to 48 numbers of ships can be maintained at 12 knot. The beam of ship is related to the width of channel, PIANC recommend three times beam for sizing of channel width. With this risk definition for width of the channel can be provided.

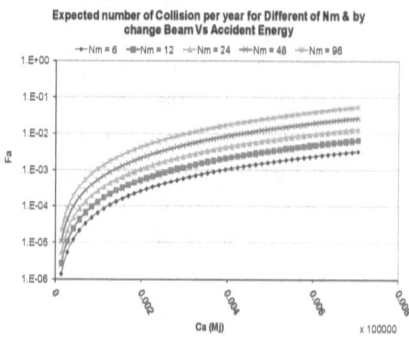

**Figure 4.22:** Risk and beam of ship

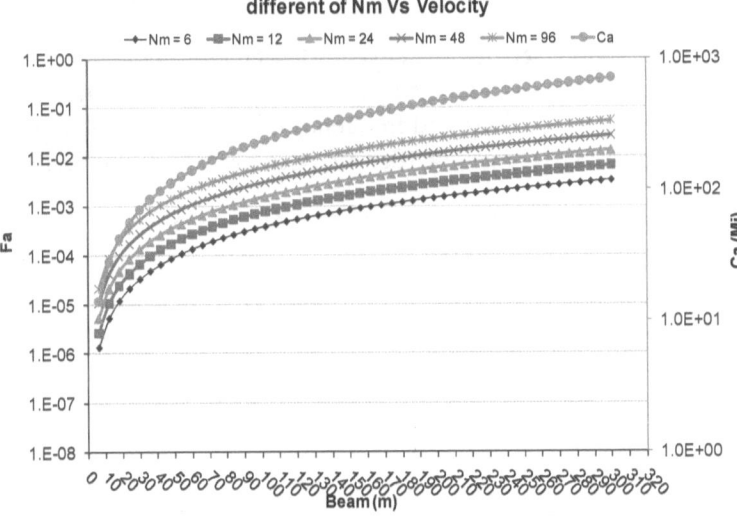

**Figure 4.23:** Risk , consequence and beam of ship

## 4.6.5 Accident Energy and Frequency at Angle of Collision

Figure 4.24 show risks and variation of course angle; angle of collision has no much impact on the risk. Figure 4.24a also revealed that risks and variation of course angle; angle of collision has no much impact on the risk. Figure 4.24 b shows that above 400MJ release of energy, higher risk is around 50, 70 and 90 degree of navigating angle. Collision impact is observed to be far more intense at angle of 22 degree.

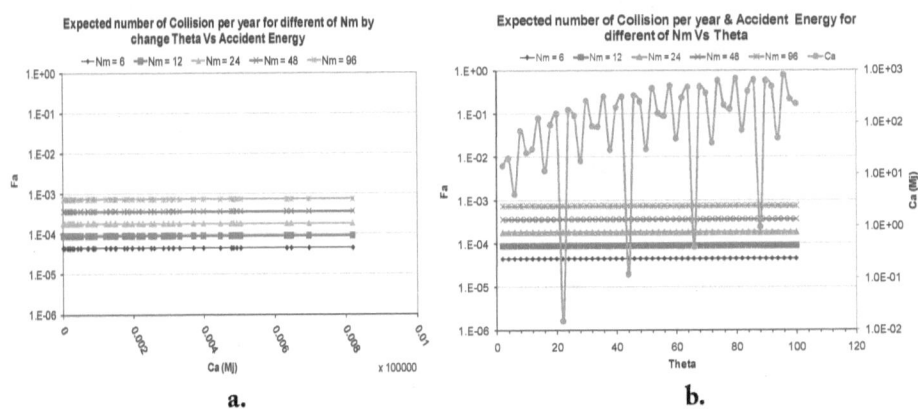

a.

b.

**Figure 4.24:** Accident frequency, consequence and changing course angle

### 4.6.6 Accident Energy and Frequency at Changing Mass

Figure 4.25 shows accident energy and accident frequency at changing mass. Figure 4.25 shows accident energy and accident frequency at changing mass. Accident energy of 20MJ and frequency of 10e-3, mass of approximately 18, 000t is tolerable. Vessel mass of 10,000t, pose highest risk for collision.

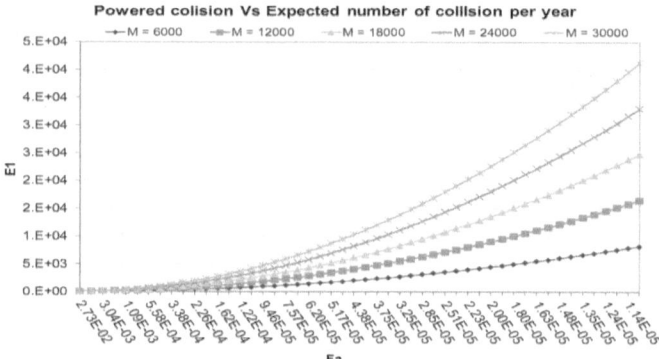

**Figure 4.25:** Accident energy and accident frequency at changing mass

### 4.6.7 ALARP Graph for Accident Frequency and Consequence Volume of Collapse and length of collapse

Figure 4.26 show the length of collapse for diferent mass of vesel and respective regresion equation and correlaton. Figure 4.26 a, b, and c show the length of collapse and Vc for diferent Nm of vesel and volume of collapse. The graph revealed that length of collapse is 7.5m at accident frequency of 10E-3.

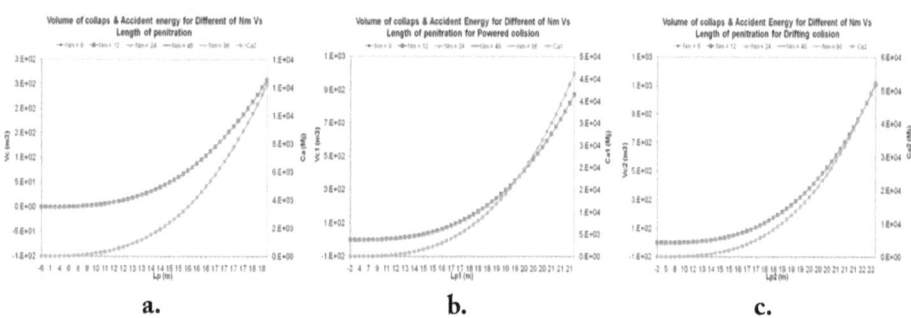

|     |     |     |
|:---:|:---:|:---:|
| a.  | b.  | c.  |

**Figure 4.26:** a. Frequency and damage (volume of collapse) and b. the length of collapse graph.

## 4.6.8 Combine Channel Variables and Risk

Figure 4.27 represent the combine plot of most variable of consequence analysis. The cross plot shown in Figure 4.27a and b represent the combine plot of most variable of consequence analysis. The interaction of the channel and vessel parameters is observed. Figure 4.27a and b shows combine plot for accident frequency variable. There is interaction between B and V at acceptable accident frequency acceptability between 10e-4 to 10e-3 and energy release of about 100MJ, and also the interaction between B and Nm shows that that at frequency acceptability of 10e-4, about 100-150MJ. The graph shows that the variable fall within the ALARP region. The he combine plot for other consequence revealed that vessel speed poses more influence on volume of collapse and length of collapse.

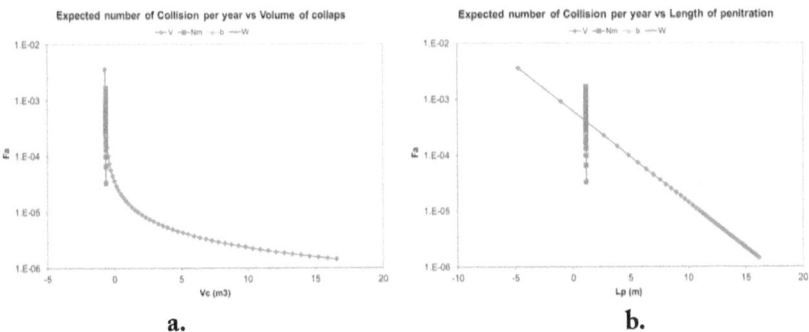

a.  b.

**Figure 4.27:** Accident frequency and volume of collapse

## 4.6.9 Alternative Consequence Quantification

Table 4.4 shows result of consequence for 96 numbers of ships. Table 4.4 shows maximum alternative consequence for the largest number of ship for the channel.

**Table 4.4:** Alternative consequence quantification

**Table 4.4:** Alternative consequence quantification

| F | M | OS | | CO2 | | RC | SRA | RA/Y |
|---|---|---|---|---|---|---|---|---|
| 2.00 | 3,000,000 | 1.4~4.4 | 4400 | 3~30 t/y | 30 | RC 2 | 0 | 0 |
| 3.00 | 30,000,000 | 44~144 | 144000 | 30~300 t/y | 300 | RC 3 | Negligible | Negligible |
| 4.00 | 30,000,000 | 44~144 | 144000 | 30~300 t/y | 300 | RC 4 | Minor | Negligible |
| 5.00 | 30,000,000 | 44~144 | 144000 | 30~300 t/y | 300 | RC 5 | Significant | Tolerable |
| 6.00 | 30,000,000 | 44~144 | 144000 | 30~300 t/y | 300 | RC 6 | Serious | unwanted |
| 7.00 | 30,000,000 | 44~144 | 144000 | 30~300 t/y | 300 | RC 7 | Severe | Unacceptable |
| 8.00 | 30,000,000 | 44~144 | 144000 | 30~300 t/y | 300 | RC 8 | Catastrophic | Unacceptable |
| 9.00 | 30,000,000 | 44~144 | 144000 | 30~300 t/y | 300 | RC 9 | Disastrous | Uneptable |

### 4.6.9.1 Result of Fatality

Figure 4.28 a and b show the agreement of frequency and consequence and corresponding number of fatality. Figure 4.28 shows the two dimensional plot of frequency and consequence and corresponding number of fatality. The two dimensional ALARP plots of accident frequency, consequence and fatality, fatality of 1 can start to be experienced as number of ship increase to 24 ships.

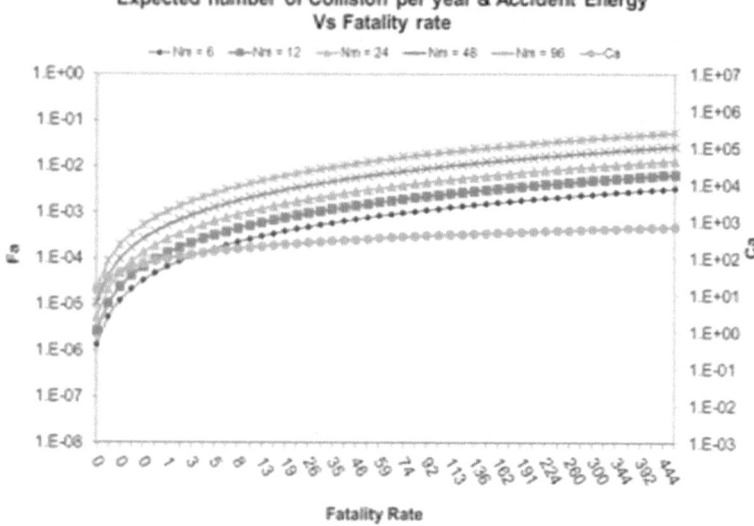

**Figure 4.28:** Risk of Fatality

### 4.6.9.2 Result of Oil Spill and COx

Figure 4.29 shows equivalent oil spill that can result for above prediction. Figure 4.33 shows cost of Cox release. Figure 4.29 also shows domains of experienced fatalities and costs versus annual occurrence probability for different types of engineering structures use as point of reference for measurement of outcome related to cost and other consequence transformation. Figure 4.29 shows corresponding oil spill that can be release as a result of accident risk. Figure 4.29a show that high number of ship and width of channel problem impact oil spill. Figure 4.30c shows that reckless high number of ships can lead to unacceptable release COx (1e+6 t /year). Figure 4.30b and 4.30c show that number of ship at 24, 48 and 96 have long term COx release impact, where as other parameters such V and beam contribute less 4. shows that the agreement of 1.5 mil with 296 MJ release of energy.

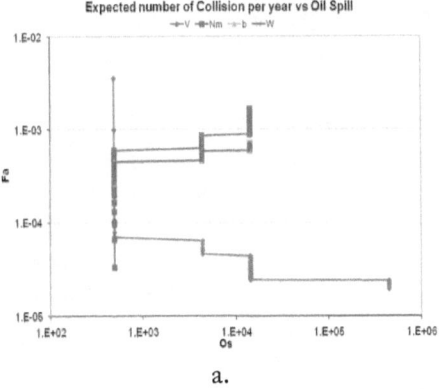

a.

**Figure 4.29:** a. Oil Spill

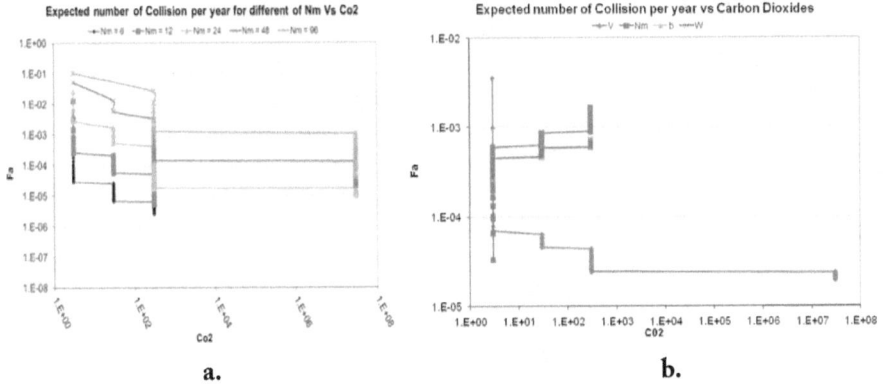

a.                                               b.

**Figure 4.30:** b. & c.Cox

## 4.7 Result of Reliability

Section 4.9.1 and 4.9.2 shows frequency and consequence test, and section 4.33 shows ALARP test. Reliability analysis is required to have assurance on the model; the purpose of reliability testing is to provide confidence and detect potential detect problems with the design as early as possible. The analysis of reliability on the model provides confidence that the system meets its reliability requirements. Reliability testing is performed at system and subsystem level as required. Accident average and projection rates per year are projection for the model to obtain mean, standard deviation, poison distribution; binomial distribution that provide predictive trend and information on system risk, capability and failure status. Minitab is employed to test the model reliability. The results are discussed in the following section.

### 4.7.1 Result of Frequency Analysis

Figure 4.31 show Fa capability test. Figure 4.31a, 4.31b, 4.31c, 4.31d and 4.31e show the overall capability for Fa, at 95% confident, only 13% of the process is out of specification, the process is stable as p value is less than 0.005 and the bell shape distribution fit well with only 1 outlier. All the representation of the data is within limit.

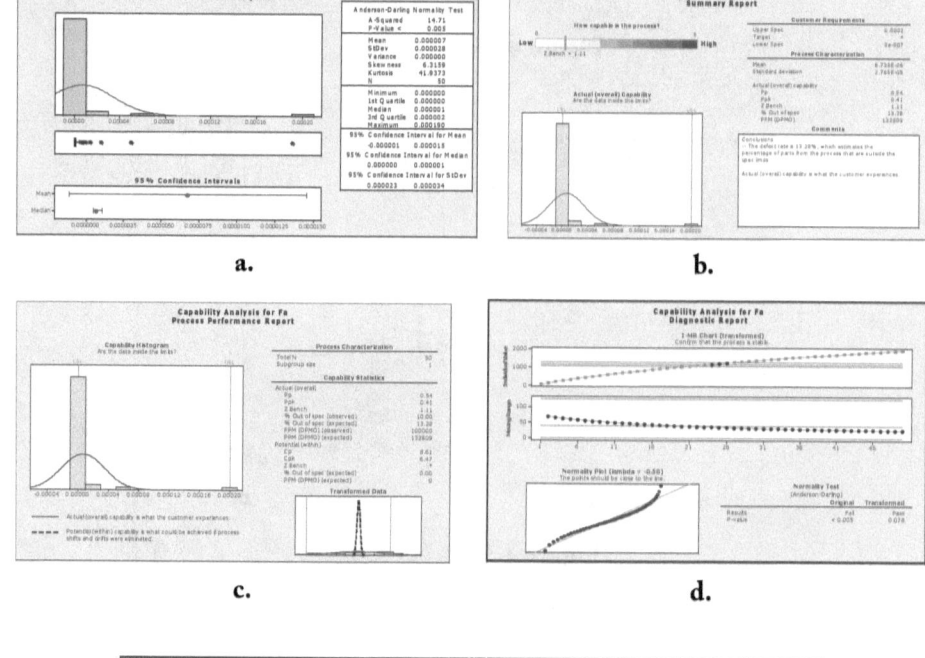

**Figure 4.31:** Accident frequency capability

Figure 4.32 and 4.33 show accident frequency residual plot with good fitness. Figure 4.32 shows accident frequency, the residual. Fitted value has good fitness; it comes with only 1 single outlier. In Figure 4.33 the normal plot shows closeness to the trend line, it shows good reliability for the data with one outlier.

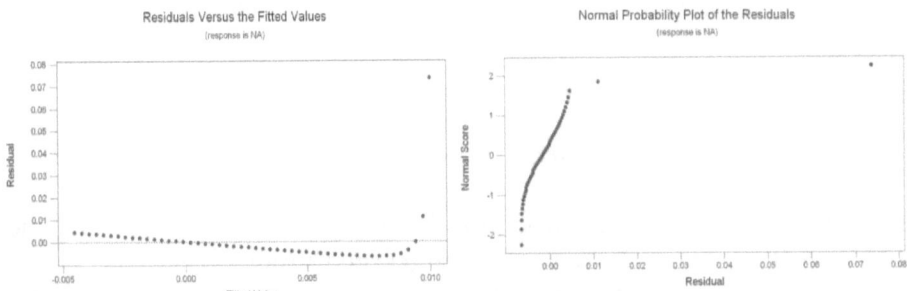

**Figure 4.32:** Accident frequency residual plot **Figure 4.33:** Accident consequence validation

Figure 4.34 shows residual histograms distribution diagram for accident frequency, skewed to low risk area, outlier can be removed. Figure 4.34 shows residual histograms distribution diagram for accident frequency, the plot skewed to low risk area show good trend for accident frequency, outlier can be removed for complete data form.

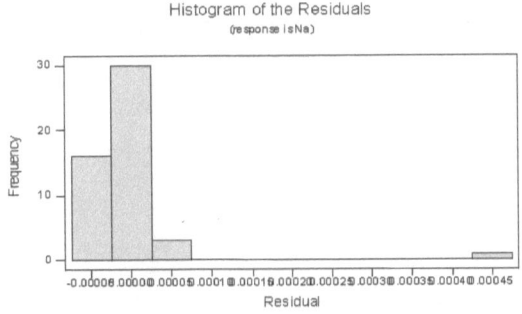

**Figure 4.34:** Residual histograms distribution diagram for accident frequency

Figure 4.35 shows Log normal plot Accident frequency (Na). Figure 4.35a shows Log normal plot for Accident frequency (Fa), the distribution shows a good to fit curve. The Figure 4.35b shows normal plot for accident frequency (Fa), with Nm change. Log normal plot for accident frequency (Fa), distribution lies between good to fit, as the lines aligned well.

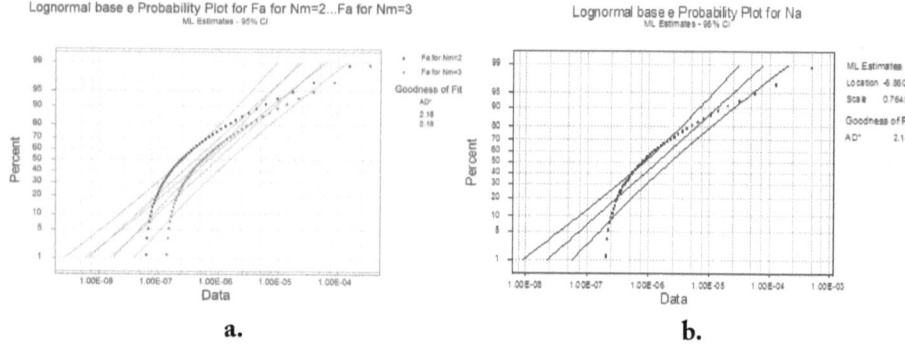

a.                                        b.

**Figure 4.35:** Lognormal plot

Figure 4.36 shows residual, probability, distribution and normal plot diagnosis at 95% confidant, accident target, and mean 10E-4. The Figures show a good curve to fit reliability. Figure 4.36 show residual, probability, distribution and normal capability plot at 95% confidence, at accident target, and mean 10E-4, binomial distribution for event that occurs with constant probability P on each trail, the likelihood of observing k event in N trail. Figure 4.36a shows residual model diagnosis plot at 95% confidant. Good reliability with good normal plot, and histogram structure. Figure 4.36b shows good correlation coefficient for weibull (0.8), lognormal (0.9) and log logistic (0.9) correlation. Figure 4.36c show good skew for Fa at 95 %confidence with one or two out layer with good correlations with mean value at 1e-5, and standard deviation 1e-1, p at 0.005.

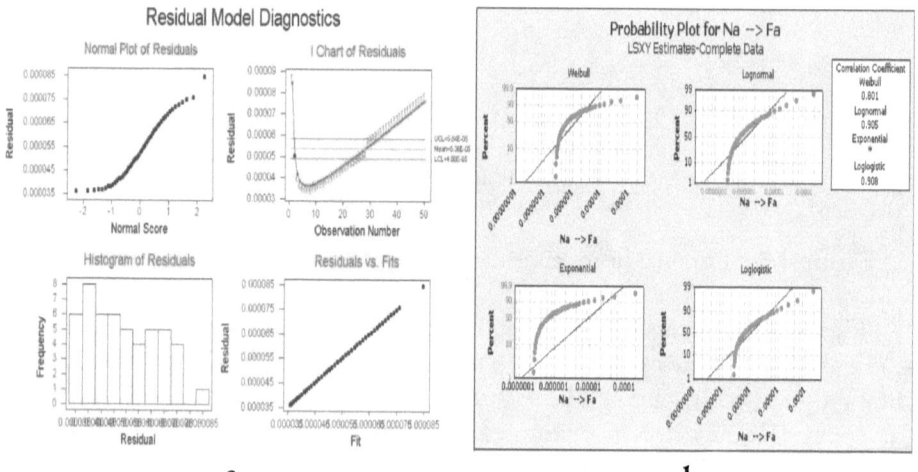

a.                                        b.

**Figure 4.36:** Normal plot

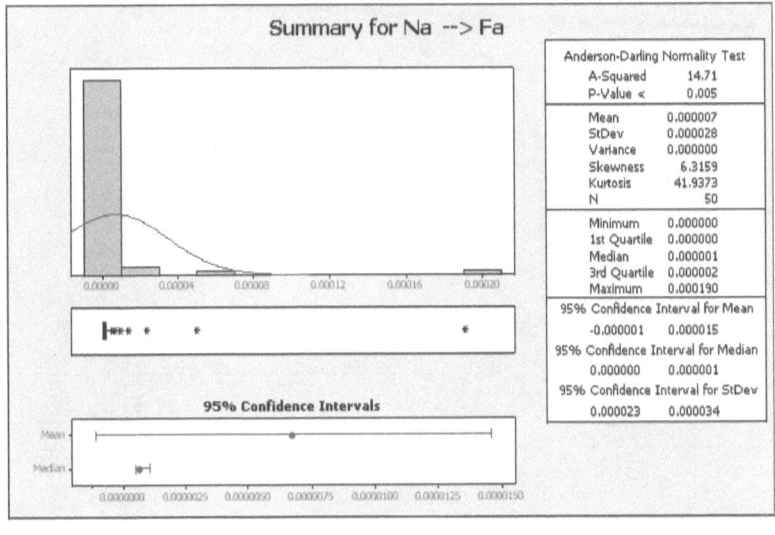

c.

**Figure 4.36:** Summary of Fa

Figure 4.37 a, b, c, d and e shows result of reliability test for Nm. Figure 4.37 shows frequency trend at varying the number of ships in the channel. Figure 4.37 a shows the Weibull process capability distribution plot with less defect and good distribution. Figure 4.37d and e also shows that the probability plot with good curve to fit as p remain below 0.005 Figure 4.37 d and e show capability report with confirmation that the process is stable, a good reliability point for the predicted values.

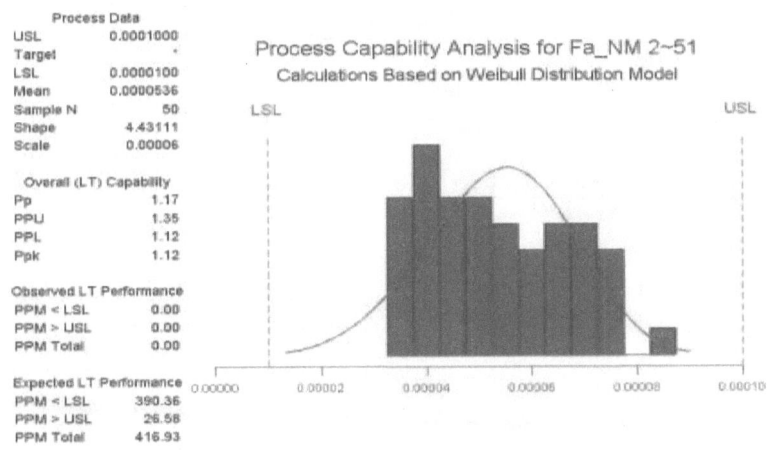

a

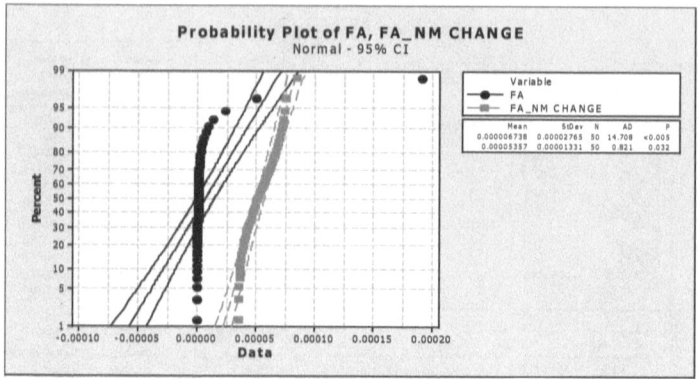

b.

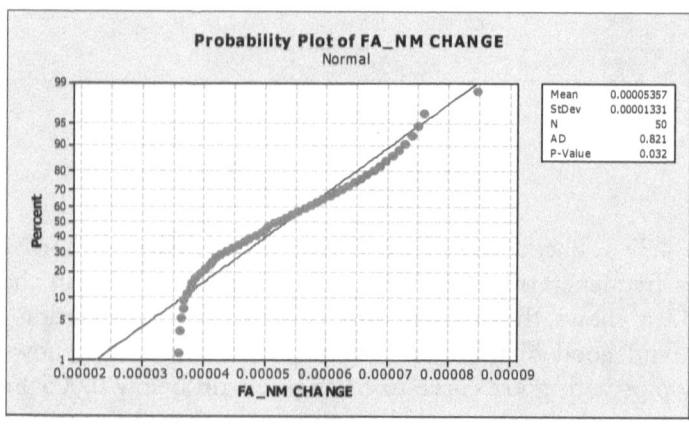

c

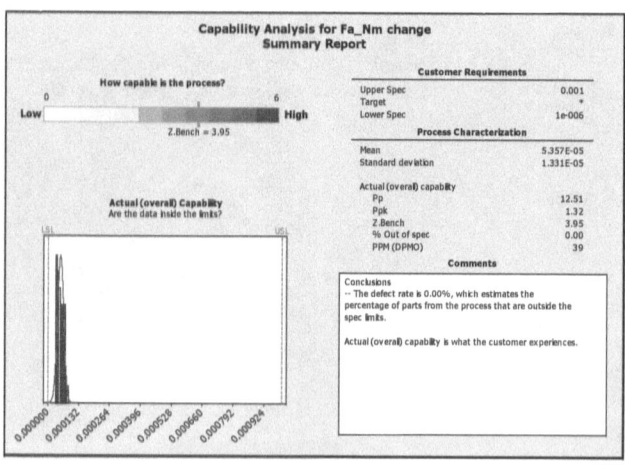

d.

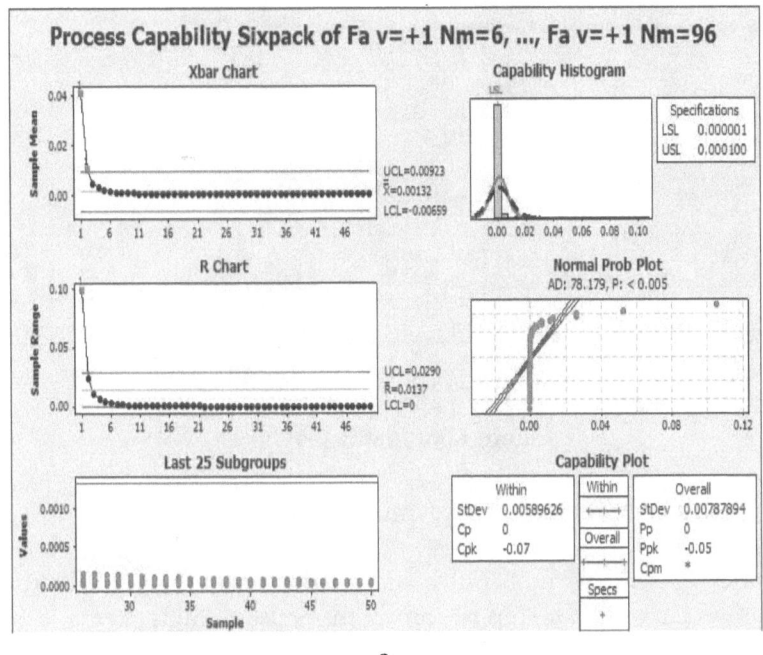

e.

**Figure 4.37:** Nm reliability

Figure 4.38 shows Matrix Plot for Risk Variables. Figure 4.38 a and b shows the matrix plot of Fa, operating region of channel variable shown in the plot is in agreement with operational risk analysis of the model. Operating speed of vessel is between 0 to 40, while operating width of the channel is between 50° to 100° and the beam is between 1° 0 to 30°. The matrix observation also shows that B and W parameter risk are tolerable for long-term.

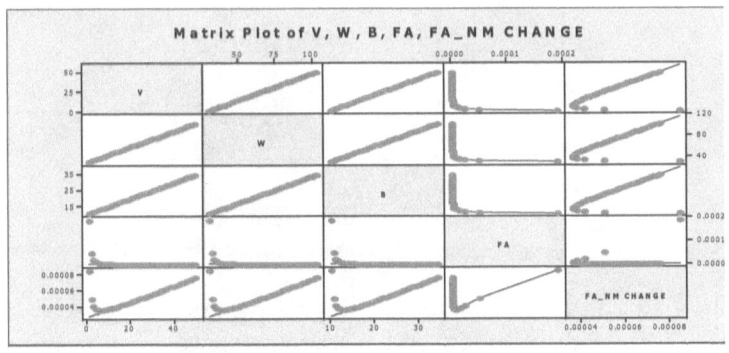

a.

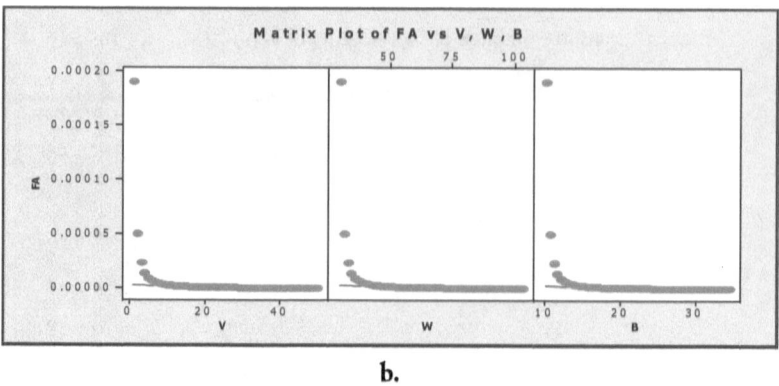

b.

**Figure 4.38:** Matrix plot for Fa

## 4.7.2 Result of Consequence Analysis Validation

Figure 4.39 shows probability and residual plot for Ca. Figure 4.39 a and b shows curve fit for ship to ship energy release, high energy is expected to tail infinity. The residual and the normal plot show good to fit curve and exponential residual. The curve also fit well. Figure 4.39a shows good p value, while Figure 4.39b show good curve fit for normal histogram configuration and form with p value less than 0.005.

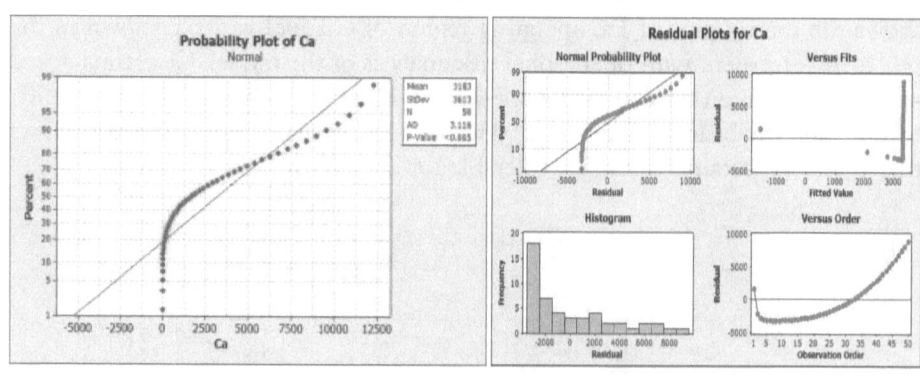

a.                                            b.

**Figure 4.39:** Probability and residual plot for Ca.

Figure 4.40 a and b show collision impact variation for powered collision energy E1, normal distribution and histogram curve. Figure 4.40 a and b show collision impact variation and response for powered collision energy

E1, normal distribution and histogram shows good trend and curve for reliability. The damage distribution is acceptable. Collision impact variation for powered collision energy E1, normal distribution and histogram curve fit is good.

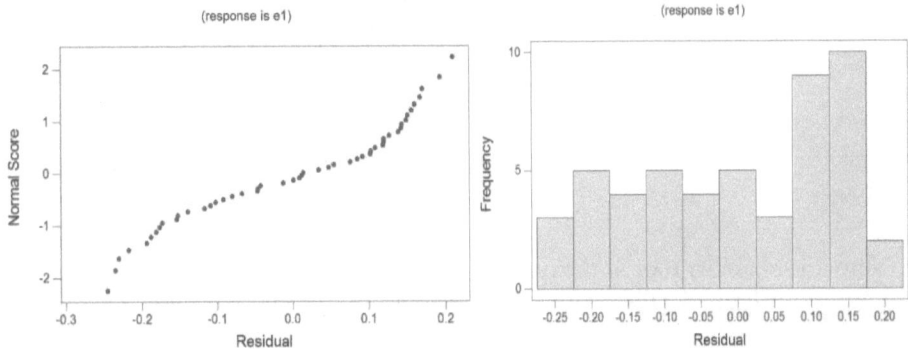

a. Normal distribution plot for E1 residual    b. Histogram plot of the E1 residual

**Figure 4.40:** Collision damage impact (Powered collision)

Figure 4.41 a and b show collision impact variation for drift collision (anchored ship). Figure 4.41 shows residual histograms distribution diagram for accident consequence (E2) for drifting collision, more distribution fall between around low risk area. Collision impact variation for drift collision (anchored ship) shows more instability compare to powered collision. Figure 4.40a and 4.41a show theoretical PDF and CDF plots for sampled data. The CDF tailing assymptote, while the PDF has exponential droping are observation of good reliability trend for the model. The overall pattern of the residual for Ca observed and plot of histogram of normally distributed show good trend for density based energy data, the "S" shaped curve on the normal distribution graph give indication of bimodal distribution of residuals. Accident consequence test, accident consequence is good to fit, residual graph of CDF profile tracing infinity.

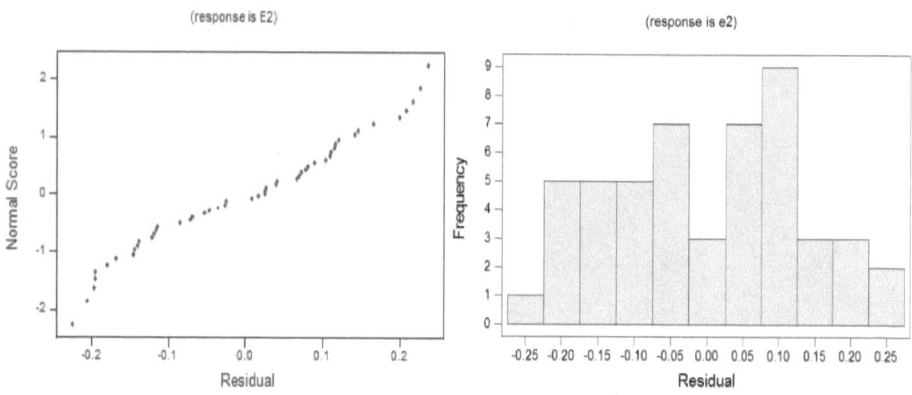

a. Nomal Probability plot of the residual    b. Histogram plot of the residual
**Figure 4.41:** Collision damage impact (Drifting collision)

Figure 4.42 shows capability plot for Ca. Figure 4.42 a, b, c, and d show the capability plot for for Ca, the report confirm acceptable result for stable process. Figure 4.42 show summary of capability for Ca, only 17% out of specification, the standard deviation and mean are minimal and the normal plot fit well. Response to energy distribution is well distributed.

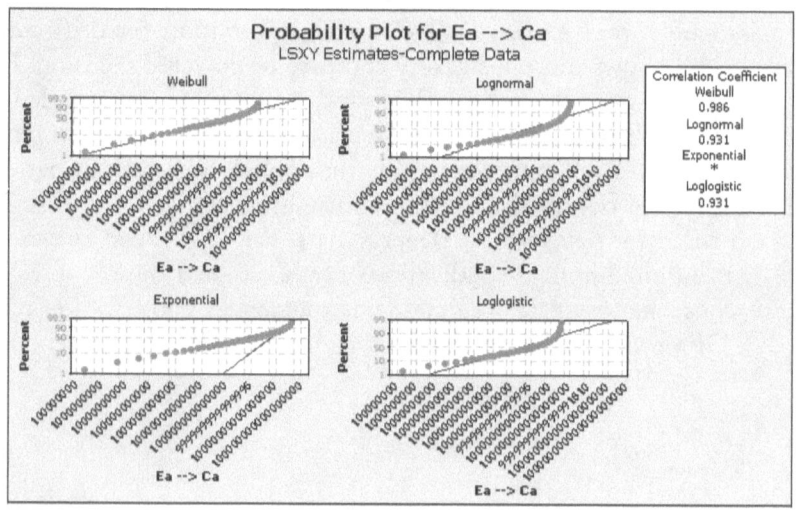

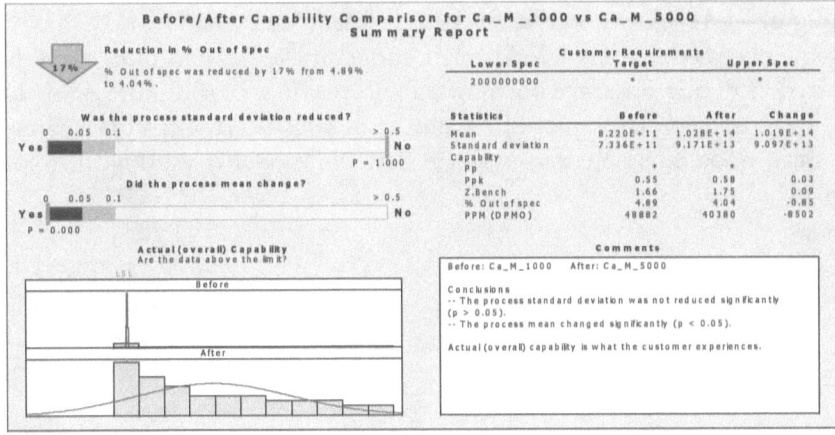

b.

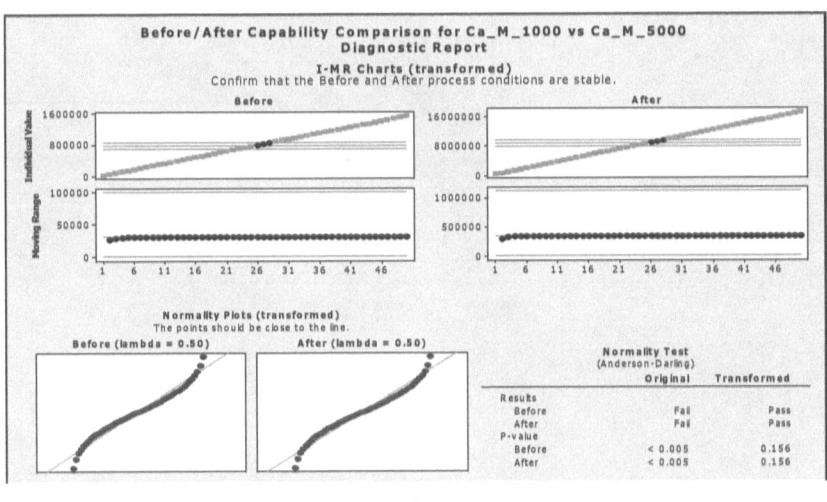

c.

**Figure 4.42:** Capability plot for Ca

Figure 4.43a shows normal plot for interpolated accident energy. Figure 4.43b shows regression plot for E1 and Vc1. Figure 4.44 shows scatter plot for Ca and Lp. Figure 4.43b shows the regression plot with standard deviation of 0.00115 and correlation up to 86. 4.43c shows normal plot for interpolated accident energy, curve fit is good. Figure 4.44 shows good curve fit for Lp scattered curve.

Figure 4.44 shows curve fit and residual analysis for frequency and consequence risk analysis. Good to fit curve and regression plot, curve fit in expected region is observed for normal and residual plot. Figure 4.44 shows good to fit for regression plot of Fa and Ea as well as E1 and Vc1. Regression plot show good curve fit that indicate that the model is valid with expected form.

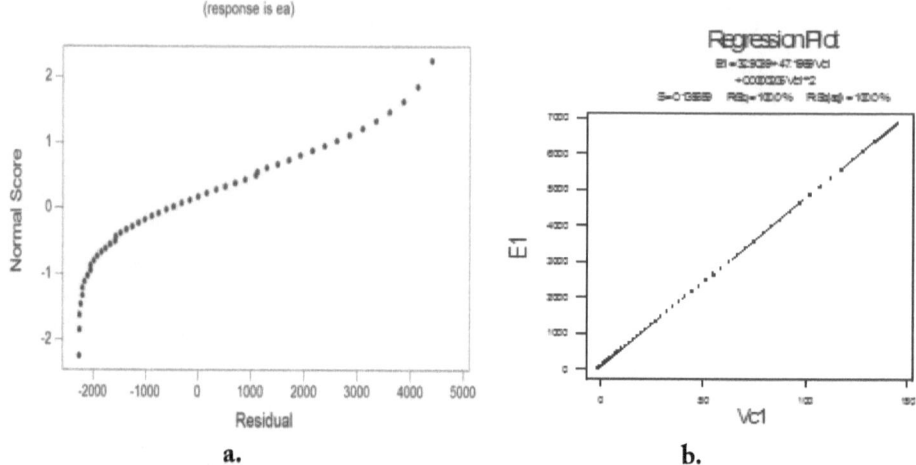

a.                                     b.

**Figure 4.43:** Normal probability plot .Ca and Lp

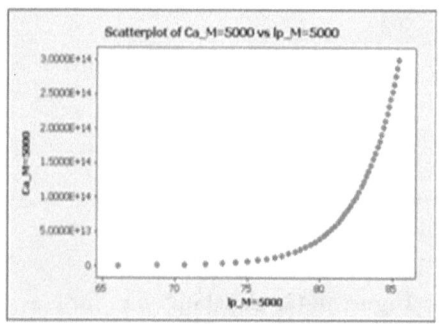

**Figure 4.44:** Scatter plot for E1and Vc1

## 4.7.3 Result of ALARP Risk Curve Validation

Figure 4.45a and b shows regression plot frequency and consequence risk model on Langat River.

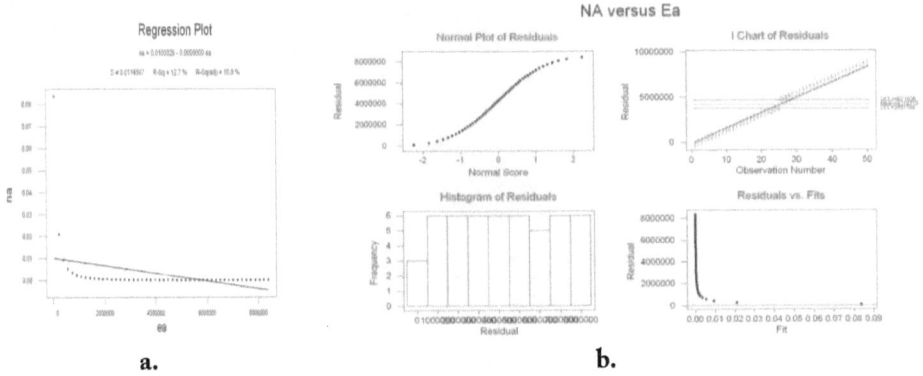

**Figure 4.45 a and b:** Curve fit and residual analysis
for frequency and consequence risk

Figure 4.46 shows Matrix plot for Accident frequency and consequence. Figure 4.46a and b shows matrix plot for accident frequency and consequence, the graphs shows of response with good pattern of inter relationships among the predictor variables. It shows that there is no much gaps and outlying in the data points. From the matrix curve it is observe that risk is totally unacceptable for reckless increase in number of ship and also high speed. The matrix observation also show that B and W parameter risk are tolerable for long-term. The matrix plot for accident frequency and consequence shows that curve fit quite well for multiple risk parameters. Figure 4.46c, 4.46d and 4.46e also revealed good curve fit matrix for operating point of the variables.

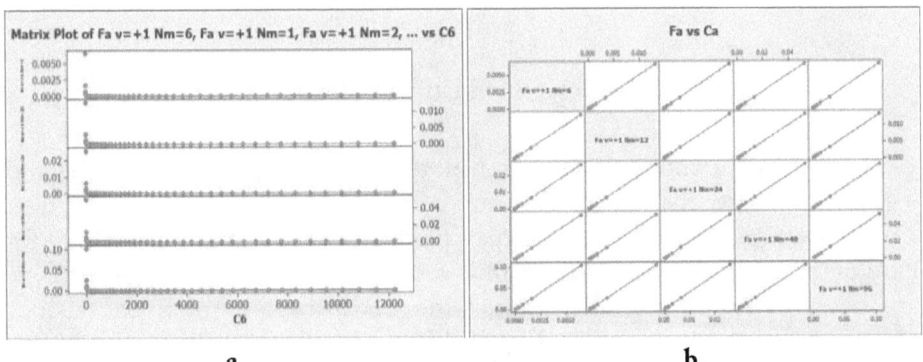

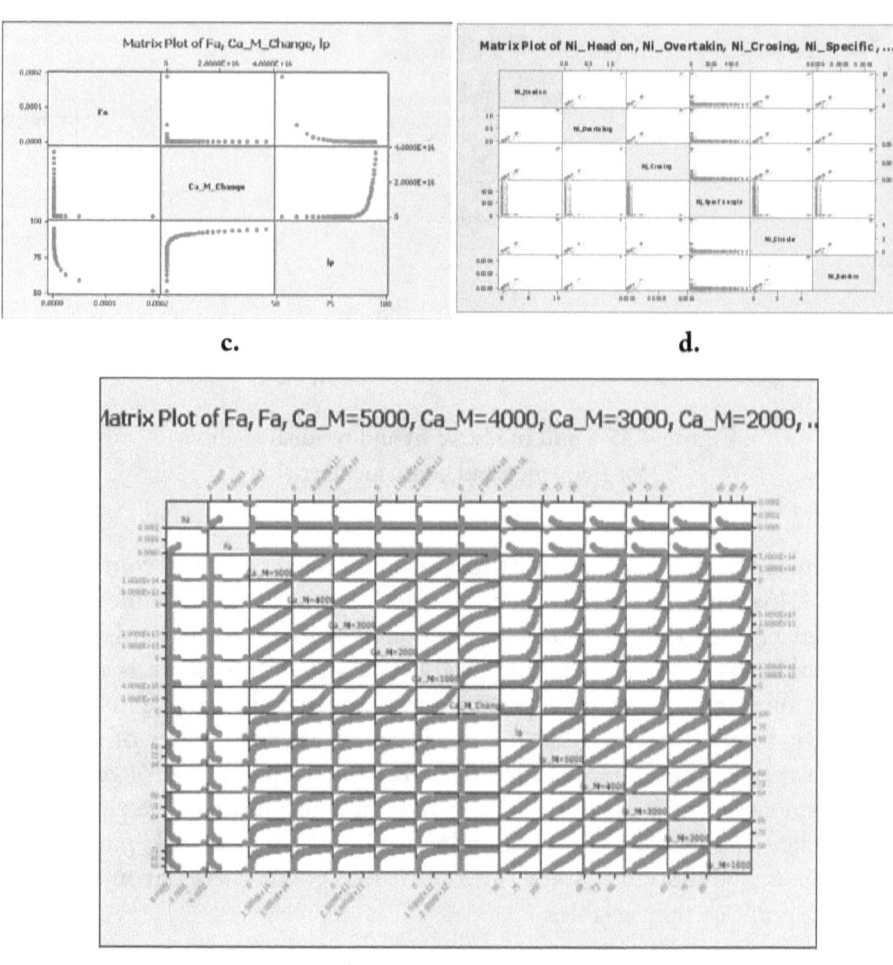

c.

d.

e.

**Figure 4.46 a,b,c,d,e:** Matrix plot for Fa and Ca

Figure 4.47 a and b show scatter plot Fa, Ca and Lp. Figure 4.47 a and 4.47b and c show the scatter plot that run through a good to fit trend line. This show a good sign for reliability of prediction, the scatter plot fit well for Fa. Figure 4.47b show the scatter plot with good distribution observed. The graphs shows scatter plot of response with good pattern of inter relationships among the predictor variables. Figure 4.8a and b shows probability plot for Ca and Cost, good correlation is obtained with weibull (0.9), lognormal (0.8), log logic (0.9), ca and CURR are not expected to exponential.

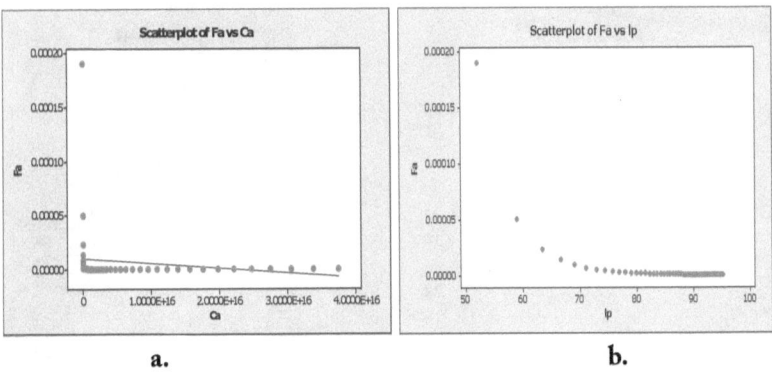

a.                                    b.

**Figure 4.47 a and b:** Scatter plot for Risk

Figure 4.48 a shows contour plot for Ni. Figure 4.48 b,c, show probability plot for Ea and CURR. Figure 448d shows time series for CURR, Figure 4.48e shows cross correlation plot for Fa, Ca and . Figure 4.48 f shows proability plot for Fa, Ca,CURR and Figure 4.48g shows surface plot for Fa, Ca and CURR. Figure 4.48d and c show good curve fit for time series curve for cost. Cross correlation between Fa and Ca in 4.48e also show good fit. Figure 4.48a shows contour plot for impact probability for random collision, there is less intensity for risk occurrence. Low risk area is less according the concentration of contour. Probability plot and surface plot for Fa, Ca and cost is shown in Figure 4.48f and Figure 4.48g show good agreement within risk acceptability of Fa, Ca and CURR. The matrix plot for accident frequency and consequence shows that curve fit quite well in low risk areas. It shows that there is no much gaps and outlying in the data points. Areas of the graph where the variables are not represent area of high risk for the variable.

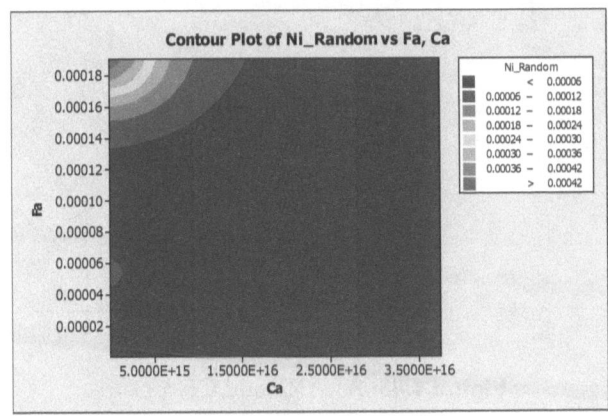

c.

211

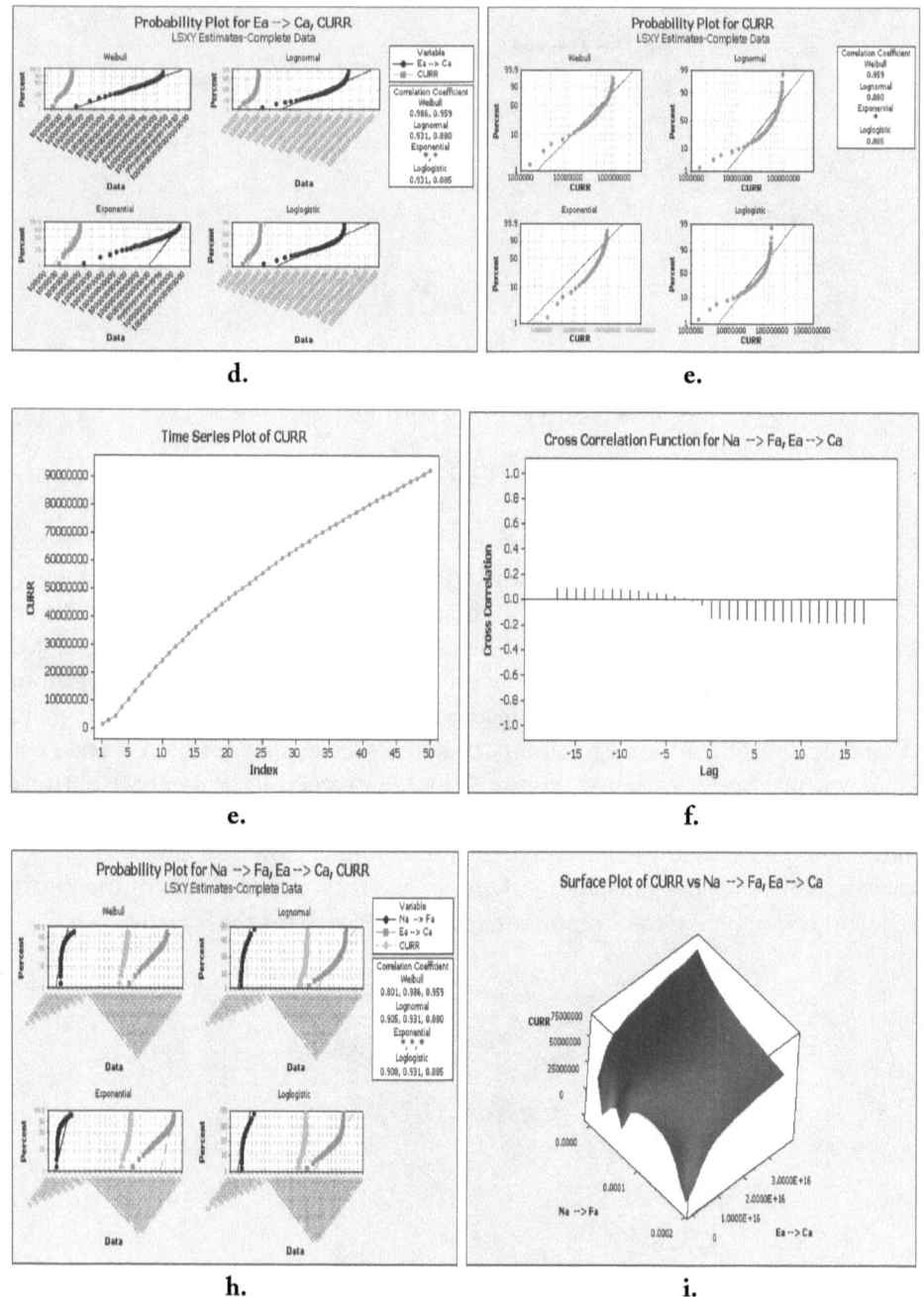

**Figure 4.48:** ALARP and CEA test

## 4.8 Result of CEA

The results for cost effectiveness analysis are in the following section.

### 4.8.1 Cost Benefit Analysis

Figure 4.49a and b shows Cost effectiveness analysis. Figure 4.49 and 4.50 shows cross plotting for accident in IWT cost variables. Figure 4.49 shows CEA result. In Figure 4.49 a and b, cost agreement with reliability and safety occur above cost of 500MJ, a figure that is in close to maximum release of energy of above 400MJ.

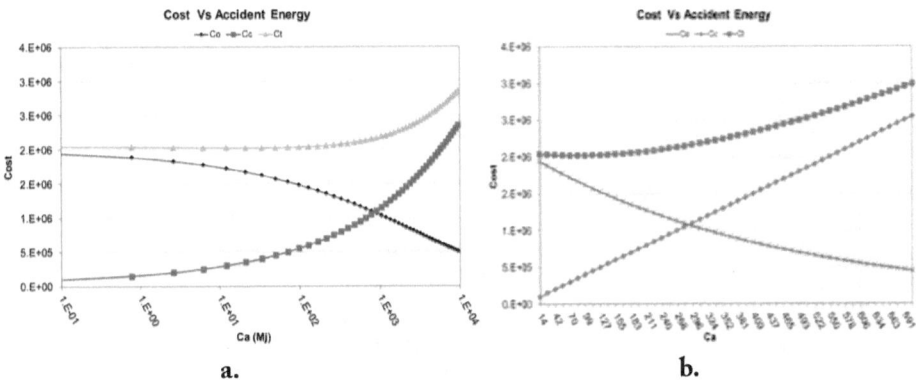

a.                    b.

**Figure 4.49:** Cost effectiveness analysis

### 4.8.2 Cost of Averting Fatality (CAF) for Velocity

Figure 4.50a and b shows cost interaction of for velocity. The corresponding cost transformation is between 2-3 million, In Figure 4.50 a and 4.50b, GCAF and ICAF show a very interesting result. The cost agreement shows that unit risk reduction of 400MJ consequence energy cost can be limited to 10,000USD. In Figure 4.50a, it is observed that 1 million USD is required to buy speed of 40 knot in the channel. In Figure 4.50b, it is observed that GCAF changing the velocity by 1 knot increment and observing the number of ship in the channel decrease.

Figure 4.51 a-d shows the cost of averting fatality interaction. In Figure 4.51a and 4.51b, it is observed that at maximum acceptability of 400MJ release of energy, 12 number of ship 3million GCAF acceptability, for minimum release of energy of 50MJ, 24 numbers of ships. The average acceptable number of ships that leads to acceptable of release of energy at 3 million GCAF is between 12 to 24. Similar is observed for NCAF but at much lower cost. CURR is much lower. ICAF describe cost related to translate consequence to fatality. Except that the amount of energy at the same cost and number of ship is lower (Figure 4.51c and 4.51d).

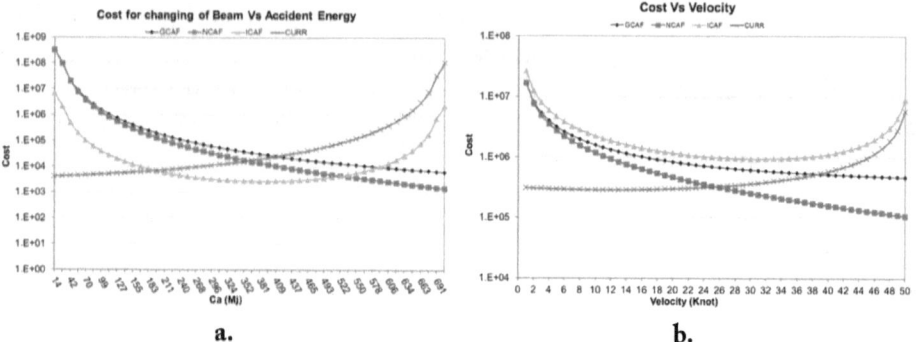

a.   b.

**Figure 4.50:** Cost interaction for velocity

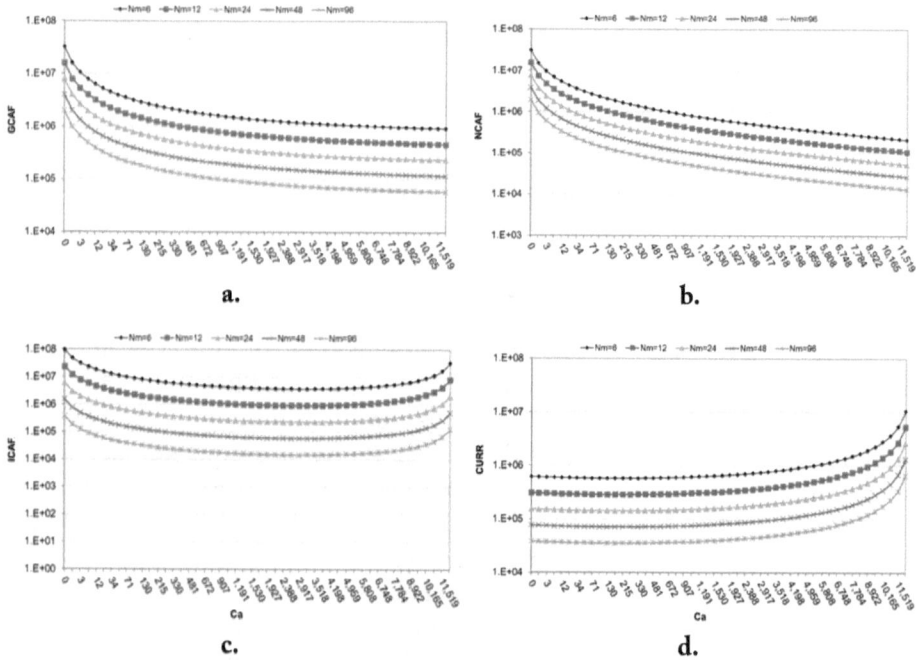

a.   b.

c.   d.

**Figure 4.51:** Cost of averting fatality interaction

## 4.8.3 Cost of Averting Fatality (CAF) for Beam

Figure 4.52a-d show CAF for beam. Similarly it trend is observed for beam increment in Figure 4.52a, 4.52b, 4.52c and 4.52d. It is observed from the plots that lower number of ships attract high cost for GCAF, NCAF and CURR, while the opposite apply to ICAF. Table 3.16 and Figure 3.27a shows dynamic of consequence translation to oil spill and COx release from cost and energy analysis, cost for impact from ship accidents, the combination of direct and indirect costs to the ship owner and other affected parties e.g. ports, governments etc and intangible costs to people and the environment are ultimately passed back to society. Figure 3.27b show measure of cost of reliability. All parameters converge at maximum cost unacceptability of RM 10 million. This is conservatively acceptable for waterway that really means to go beyond compliance.

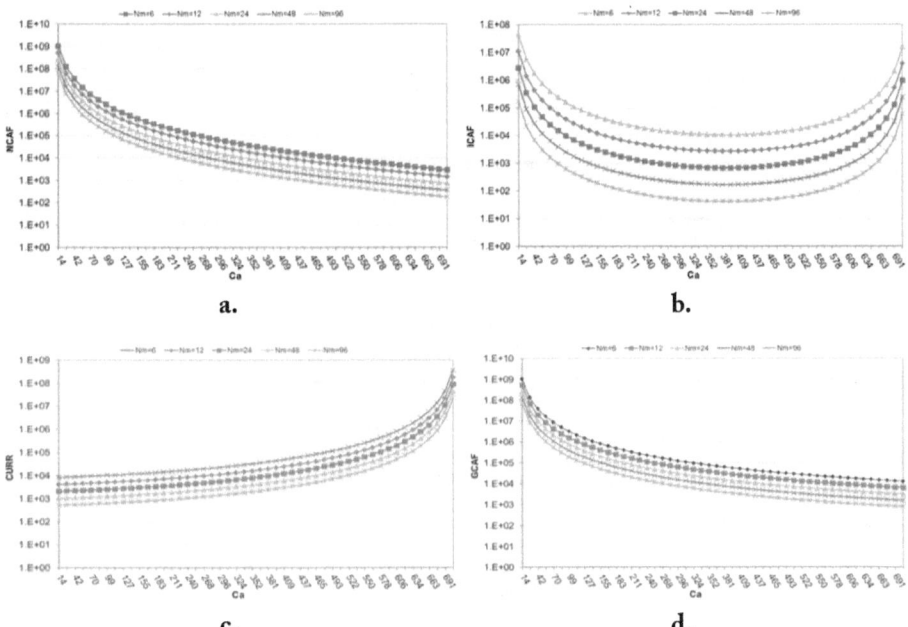

**Figure 4.52:** CAF for beam

### 4.8.4 Cost of Alternative Consequence Transformation (oil spill and GHG)

Figure 4.53 shows cost for combine interaction of accident variables. Figure 4.53 shows cost agreement for all variables with maximum at 10 million RM, at acceptability of 1e-3, the cost is between 3-5 Million, a good value of agreements.

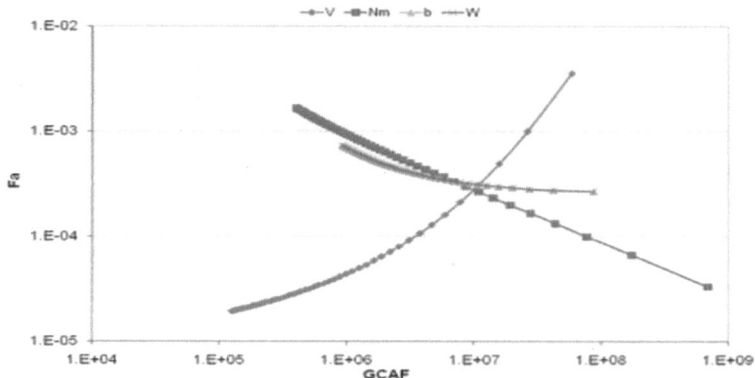

**Figure 4.53:** Combine cost plot for accident variables in IWT variables

## 4.9 Result of RCO

Determination of risk associated with channel under the Nm, V, B, W, M parameters leads to RCO grouping into major collision contributing factors. Table 4.4 shows RCO grouping from waterways risk.

**Table 4.5:** RCO

| RCO | Description | PLL | Benefit | GCAF (Mil) |
|---|---|---|---|---|
| Loss of navigation function | V | 1 | 0 | 2.5 |
| | W | 0 | 1 | 1.5 |
| | Nm | 0 | 1 | 1 |
| Loss of propulsion function | V | 1 | 0 | 2.5 |
| Loss of mooring | M | 1 | 0 | 2.5 |
| Human error | V | 1 | 0 | 2.5 |

Risk associated with channel under the Nm, V, B, W, M parameters are grouped to determine RCO for major collision contributing factors. Theta is not included because of its trend with fewer changes to waterway risk, thus, it is taken into consideration under uncertainty analysis of visibility. Risk acceptability is defines in difference ways, it signify the difference between the expected income the expected operational cost. The model considers added value to society at the goal based objective formulation. If the society perceives a positive net benefit, the activity may be claimed to be socially responsible. Risk acceptance criteria are is given in the form of bounds on the annual probability of failure independence of the consequence of failure which is measured as energy, monetary costs or 1 translated to lives lost or oil spill (OS), or COx. Figure 4.51, 4.52, and 4.53 shows alternative consequence deduced from Table 3.27. In the context of environmental requirement, the rationality problem of setting public acceptance criteria for the operation is essentially that there are two decision makers with partly conflicting settings of the preference by the owner and the public. The lower the CAF of a RCO, the more priority has to been given to its implementation. Table 4.4 shows summary of deduced result of interrelation between societal risk consequence acceptability.

The analysis of CEA is followed with RCO`s analysis and balancing of cost benefit and reliability towards recommendation for efficient, reliable and effective decision. RCO work involve iteration and revisiting initial condition like regulatory, baseline data, historical data, in situ data, data from other studies and functionality information. Advantage derived from channel improvement work like channel widening, deepening and straightening, navigation aid, twin propeller system is quantified into sustain equity for determination of cost control option to provide reliability and safety for IWT as required. The total risk metric is represented by $R = \Sigma$ (Fim x Cm) + $\Sigma$ (Fjm xCjm) x $\Sigma$ (Fkmn x Dkmn). Where, Fim = Annual probability of mitigatable collision, Cim = Annual probability of unmitigatable collision, Fjm = Probability of other failure mode, Cim/Cjm/Dkmn = Consequence severity.

Inland ports are mostly situated in close proximity to people. That make proactive safeguard of people and the environmental through tope down systemic risk model a very important study in development work. The general rules and regulations that govern ships at sea do not address the particular concerns of inland ports and vessels. Specific rules and regulations are developed by local authorities regarding safe margin that need be observed to avoid accident. IWT is expected to survive the normal effects of flooding following assumed waterway system and channel complexity by some external forces. Likewise vessel significant releases or reduce damage of IWT is expected to survive effect of flooding resulting from hull rupture or other structural problem as a result of accident event. Design characteristic that can help to prevent significant releases or reduce damage of IWT vessels in case of accidents are described:

217

i.   Segregation of the cargo area and accessibility to spaces
ii.  Vessel fire protection systems
iii. Cargo containment design requirements, double hull, requirements for cargo containers
iv.  Structural analysis requirements, construction and testing
v.   Secondary barrier provisions and low temperature provisions, cargo pressure and temperature control, pressure vessel and piping, pressure relief systems and vacuum protection systems
vi.  System requirements, emergency shutdown valves and shutdown systems, emergency release couplings, vessel, gas detection systems, leak detection for gas carrier hold spaces.
vii. Location of cargo tanks, cargo tank instrumentation,
viii. Gas detection and safety shutdown systems for terminal operations

Table 4.4 shows alternative consequence for the maximum number of ships in the channel. Table 4.5 show summary of RCO, V has one potential loss of life (PLL), therefore no benefit and it will required 2.5 million to rectify loss of navigation, loss of propulsion, loss of mooring and human error as a result of V Monitoring. Width and Nm affect only navigation issue; they pose less risk of PLL.

RCO for each collision situation is defined based on the above. In order to identify new RCO, generated result from the analysis of frequency and consequence, cost and benefit is weighted for effectiveness. This can be supported by with expert rating to contribute to possible risk prioritization control options for IWT of on the channel, thus this is not considered in this study as the requirement for that is replaced with reliability analysis. The descriptions of the major hazards and corresponding risk control options from the hazard identification and the results from the risk analysis which are summarized could be presented to the group of experts for further validation, in this study statistical software and case study are employed for validation. From the risk study, prioritized RCO that were selected for further evaluation in terms of cost effectiveness assessment are discussed in the next paragraph. Even thus this research is about collision, the impact to collision is not far from contact and grounding collision scenario. Therefore some of the measure that is recommended to be taken could benefit curbing accident from contact and grounding. The main RCO`S are:

i.   Improved navigational safety
ii.  Redundant propulsion system: two shaft lines

iii. Required maintenance plan for critical items as well design requirement for increase double hull width, increase double bottom depth or increase hull strength for transportation of oil in the channel

iv. Human factor and human reliability is quite critical in risk work, it is recommended to be done separately channel complexity and uncertainty is also recommended to be done separately on case by case as required.

## RCO 1: Improved Navigational Safety

Improved navigational safety can be achieved in a number of ways. From various identified risk control options, five cost effective risk control options for navigation improvement that could potentially reduce the frequency of collision and grounding which are: i) ECDIS (Electronic Chart Display and Information System), track control system, ii) AIS (Automatic Identification System) integration with radar, iii) Improved bridge design and ) Traffic Separation Scheme (TSS). The risk control options related to navigational safety in the list above might be promising alternatives for IWT. The cost effectiveness of implementing this measure for IWT is evaluated in this study. Hence, the risk control option for improved navigational safety is defined as implementation of one or more of the above alternatives. Analysis of TSS is favorable for risk reduction in IWT.

ECDIS is a navigational aid that can be an alternative to nautical paper charts and publications. It is use to plan and display the ship's route as well as to plot and monitor ship positions throughout the voyage. It is capable to determining a vessel's position in real time. The use of this aid would reduce the navigation officer's workload compared to using paper charts. Although not mandatory for IWT, the use of ECDIS onboard current vessels is quite common and the proposed risk control option is to make ECDIS mandatory for vessel plying IWT. Track control systems as part of ECDCS integral system are continuously comparing the vessel's actual course with that originally planned. The intended route is planned before departure and entered in the track control system. The real time information from navigational equipment is utilised in order to ensure that the planned route is followed. Additional risk reduction can be achieved by integrating the track control system with the ECDIS.

Automatic Identification System (AIS) is designed to send and receive information in relation to a vessel's identity, course and cargo. Current regulations require such information to be presented in an AIS display and the minimum requirements are three lines of data consisting basic information of a selected target. The AIS can also be integrated with the radar's Automatic Radar Plotting Aid (ARPA) function so that the additional data is available in the radar display. Benefits from the AIS—ARPA integration will include enhanced situation and traffic condition awareness and improved ability to make early decisions based on real-time data for avoidance of potential collisions. Other improvement of radar performance and navigation can be achieved in the following ways:

i.   Identification of ship names for radar targets and clarify the target's intentions
ii.  Detection of targets which are in radar shadow areas and extends radar range
iii. Easier access of AIS information
iv.  Accounting for the ship's rate of turn to predict accurately the target's path.

Improved bridge design, this involve compliance with SOLAS, SOLAS contains minimum requirements regarding mandatory equipment. It gives limited requirements related to bridge layout. Improved bridge design means an upgrade from standard. Minimum SOLAS bridge design provides improved performance of all navigation related tasks as well as enhanced cooperation between the bridge team members. This can enter into limiting human error as well. The goal is to enhance the navigational safety by reducing the probability of failure in bridge operations by any cause. For the purpose of this cost effectiveness assessment of the following voluntary DNV class notation for navigation safety could also be adopted: i) Design of the workspace and the bridge layout, ii) Navigational equipment, iii) Human-machine interface.

## RCO 2: Machineries, Power plant and Redundant Propulsion System

Machineries reliability, installation of valve control is recommended for design ship. Machinery failure is a significant causal factor in collision accident. Collision can be avoided if the ships had redundant propulsion or steering systems. Reduce risk of oil spill due to overfilling, malfunction of a valve or human failure among other causes. The levels of storage tanks on board must be continuously monitored since overfilling or product discharge on deck could have consequences for human life and for property. The redundant propulsion and steering system must ensure that, irrespective of the ship's loading condition, when a failure in a propulsion or steering system occurs: i) The maneuverability of the ship can be maintained, ii) A minimum speed can be maintained to keep the ship under control, iii) The ship can maintain operation with a redundant propulsion or steering system so that a vessel can ride out the storm or slow navigation in port, iv) The propulsion and steering functions are quickly re established. Redundant propulsion and steering systems can greatly reduce the risk of vessel disability subsequent loss of life or cargo and damage to the environment. Operational advantages of redundant propulsion are: i) Greater reliability, ii) Improved safety levels, iii) Higher vessel availability, iv) Protection of environment.

Cost effectiveness assessment for redundant propulsion systems will be achieved by installation of independent engines and two shaft lines. The use of all electric propulsion could be a good advantage for optional navigation mode. This would also have effect on different hull forms compared to ships with single propellers.

## RCO 3: Human Capital Development

Discussion with waterways authority on the case study revealed that only the captain's qualification and competency is being screened and regulated. It is recommended for institution to screen on certification and competency of all officers on the vessels and to undergo simulation for normalization of behavior. Accident analysis, assessment result and reliability analysis is done for watch keeping subsystem analysis using fault thee analysis. Human reliability analysis can best be tackled using DELPHY method recommended by IMO. The DELPHY method incorporates expert rating to tap information about waterways and to deduce risk index from expert and water ways user information. Issue of fatigue with mariners is important to be regulated. Although particular legislation exists regarding working and resting periods onboard, Potential problems related to cases where crews work long hours when at port leading to fatigue at a later time when the ship has left port remain of accident risk occurrence causal factors. Thus, this control option is an extension of the STCW/ILO regulations regarding rest periods to the time spent in port in order to obtain reduction of crews fatigue and hence reduced the probability of collisions, groundings and contact risk control.

This risk control option relate to HRA also aims at increasing the bridge team's ability to handle difficult maneuvering tasks and crisis situations by increased use of simulator training. The effect of such training could provide better navigational safety and a reduced risk of collision, grounding and contact events. The simulator training could be specially designed for particular port environments, underwater topography, and particular bridge layouts on specific vessels and would give the participants exercises in handling challenging situations from different positions of the bridge team. Important parts in such exercises might be passage planning, situation awareness and operation during malfunction of critical technical equipment. The risk control option suggested herein goes beyond the basic training requirements defined by IMO's International Convention on Standards of Training, Certification and Watch keeping for Seafarers (STCW). Other insignificant RCO are: oil Spill, emission, and dredging.

## 4.10 Result of Validation by Case Study

The results of validation by case study are presented in the following section. Inland water transportation operation attracts low probability and high consequence accident. This makes reliability requirement for the design and operability for safety and environmental protection very necessary. Collision is the largest accident risk scenario among water transportation risk. The study area this study is Langat River, a 220m long navigable inland waterway. Personal communication and river cruise survey revealed that the river has been under utilized and collision remains the main threat of the waterways despite less traffic in the waterways.

This make the case to concentrate on collision risk and the need to establish safety and environmental risk based model (SARM) on Langat River as case study validation for sustainable development of the IWT a necessity. Data related to historical accidents, transits, and environmental conditions were collected. Accident data are quite few, this is inherits to most waterways and that make probabilistic methods the best preliminary method to analyse the risk. The result deduced from the case study validation can be optimized through expert rating and simulation methods as required. For this case, this is not required as the study has been validated through reliability analysis which considered a better approach.

Figure 4.33 shows the Langat River tributary IWT system. Barge and tug of capacity 5000T and 2000T are currently plying this waterway at draft of 9 to 15 respectively. Collisions (including contact between two vessels and between a vessel and a fixed structure), causes of collision are linked to navigation system failure, mechanical failure and vessel motion failure. These factors and their interconnection are considered in PrHA towards use of the river for transportation. Safety associated with small craft or other object or waste signature is not taken into account in this model.

The data require from Langat River to test the model developed are provided in chapter 3. The result can be use for implementation for IWT safety and environmental risk deduced based on system engineering reliability based approach. Environmental factors considered in the Langat River case study include marine processes, particularly the tides and tidal currents. Tides in the Malacca Straits have a mean spring range of 4.3 m. Tidal currents are bi—directional, but set to the northwest, and average tidal current velocities range from 1.5 m/sec to the northwest to less than 1.0 m/sec to the southeast. This constant tidal reworking of bottom sediments in the shallow (120 to180 m) straits results in concentration of a large sand body. Those are characterized by large asymmetrical sand waves along the bottom of the Malacca Straits.

Another reason for choosing examples of Langat River for the case study is because of recent environmental pressure emanating from global warming, climate change, and population rise leading to congestion in the city and point form pollutant release. Envisaged expansion in the area further forecast potential rise of traffic in that area for a long time. Intolerable system failure on complex and dynamic system like IWT has made decision, policy and political will very hard for development of IWT. This make the need for use of scientific system based model deduce from risk and high goal based objectives couple with uncertainty analysis consideration necessary for IWT development in Langat River. This study considered the balancing system between safety, environment, risk, and uncertainty, social, economic, and technical requirement leading to sustainable development of reliable and efficient IWT. Development of IWT involve assessment of channel and vessel size, maintenance dredging, soil, air and water quality, sustainability, risk and uncertainty. Components and requirements of these factors have been discussed in chapter two, section 2.3-2.6 of this thesis.

Table 3.29 shows Langat River improvement activities that planned for channel. These remain part of risk consideration. Figure 3.31 shows Langat River channel width and dimension. Table 3.28 shows the chronology of accident along Langat River due to inland navigation activities. Table 3.19 shows approach channel, Table 3.20 shows channel straight dimension data, Table 3.32 and 3.33 shows vessel principal dimension plying Langat River. Table 3.34 to3.36 show traffic data. Table 3.37 shows accident data. Figure 3.36 a, 3.3.6b and 3.36c show the environmental parameters considered in the risk process.

In Figure 3.36a, Tide movement and current EB, Green to red: low to high speed, In Figure 3.36b, Water level Mean, water level of 40cm seasonal variation, Existing coastal environmental current in Figure 3.36c, coastal current, Avg. Speed in Spring tide 0.4—1.2 m/s, Avg. Speed in Neap 0.2—1.0 m/s. Identification of accident scenario that is significant to risk contribution considers the use of holistic qualitative risk assessment high level objective derived from system functionality and standards matching. Table 4.6 shows preliminary hazard analysis hazard related to waterway system and ship.

In the qualitative assessment done for Generic River and the Langat River, the main risk contributing factors to collision are: power transmission failure, navigation failure, vessel motion failure, human error, other external cause. For visibility, navigation is more risky at night than day time. The qualitative assessment follows generic assumption as well as input from Langat River local situation for evenly safe distribution during day and night. A review of risk assessment methodologies applicable to marine systems reiterate that the absence of data should not be used as an excuse for not taking an advantage of the added knowledge that risk assessment can provide on complex systems. Hybrid of deterministic, probabilistic and reliability approach adopted to test model accuracy is solidify its strength as tool for development of any River. Like the case of generic model, frequency analysis is estimated from traffic density, expected number of collision per passage (Ni), Necessary period for ship to pass the fairway (T). Secondary data used for the model are probability of failure per nautical and per passage for different collision situation (Pi), failure per nautical mile per passage ($\mu_c$) for different waterways use constant of probability of losing control Pc 2.5 x 10$^{-5}$

The result of Accident frequency Analysis (AFA), (Fa) observed is checked and compared with frequency acceptability criteria index shown in frequency acceptability in Table 3.9 In maritime industry, risk index acceptability is use to measure risk, where by low risk is considered 1e-5 or 1e-6 and high risk is considered at and acceptable risk is 1e-2 is consider risky, while above exponential 6 (1e-3) is considered negligible. Like the case of generic model, intermediary result, uncertainty analysis and qualitative assessment represent intermediary results of the model.

## 4.10.1 Langat River Channel and Navigation

Figure 4.54 a and b shows waterway vessel and width dimensioning

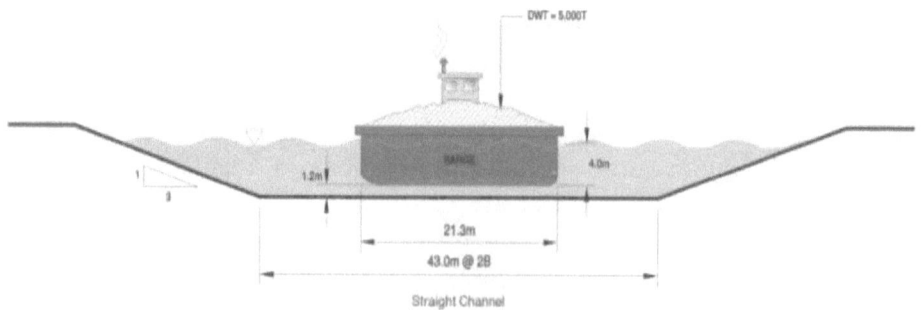

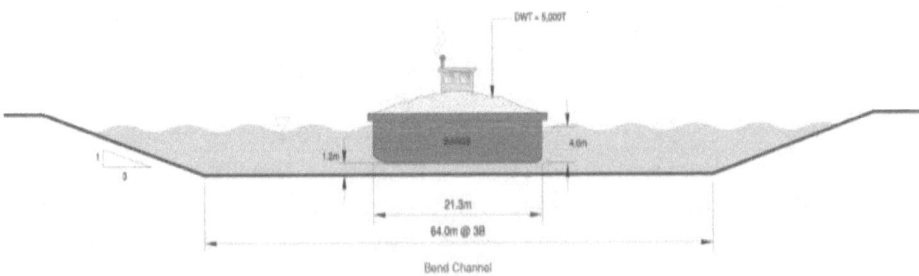

**Figure 4.54a**: Langat vessel and channel navigation parameter (2000T)]

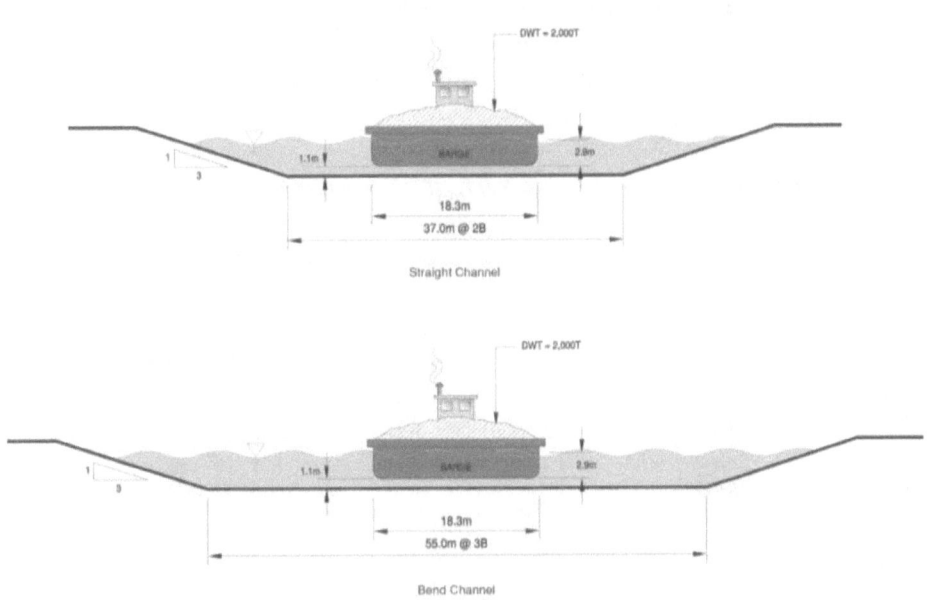

**Figure 4.54b**: River Langat vessel and channel navigation parameter (5000T)

## 4.10.2 Result of HAZID Risk Assessment

Table 4.6 shows deduce PrHA for Langat River.

**Table 4.6:** Preliminary Hazard Analysis

| Hazard element | Triggering event1 | Hazardous condition | Triggering event 2 | Potential accident | Effect | Corrective measures |
|---|---|---|---|---|---|---|
| Kinetic energy | Loss of navigation control | Ship1 sail on random course | Another ship is on ship1 course | Collision, rupture of cargo tanks | Fatalities, environmental damage, damage to hull | Improving navigational standards |

| Kinetic energy | Loss of propulsion failure control | Ship1 sail on random course | Stationary obstacle on ship 1 course | Power grounding, rupture of cargo tank | Fatalities, environmental damage, damage to hull | Use of twins propulsion |
|---|---|---|---|---|---|---|
| Kinetic energy | Obstacle on ship course | Retardation (i.e. reverse) | Movement of unfastened material on board vessel | Crushed personnel, material damage | Fatalities, environmental damage | Early detection and removal |

## 4.10.3  Result of Frequency Analysis

The following are result of accident frequency plot from the model implementation on Langat River.

### 4.10.3.1  Impact Probability and Angle of Impact

Figure 4.55 show accident impact probability correlation with speed and angle of impact. Figure 4.55 shows accident impact probability correlation with speed and angle of impact. Impact probability is likely at angle between 105 to 115 degrees. Average speed of 10 knot gives minimum impact. Impact is more like for excess speed of 12 knot at angle around 100 to 150 degree. This result is in line with the result of the generic model.

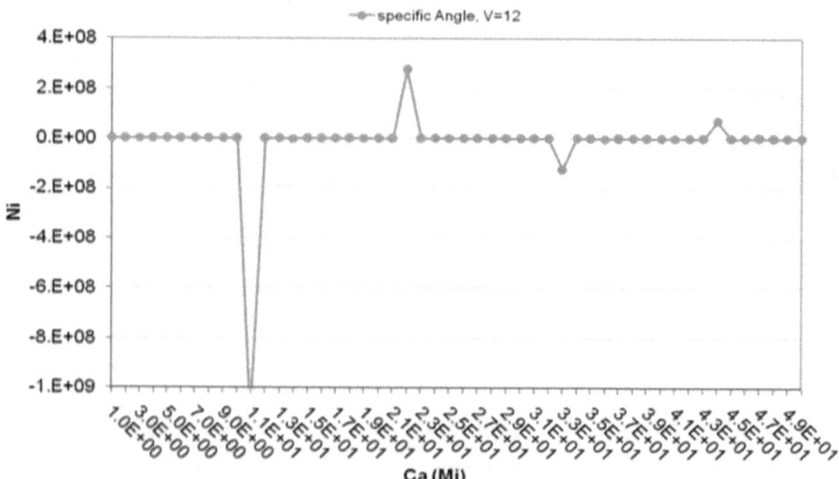

**Figure 4.55:** Accident Impact Probability and Angle of Impact @ Changing Velocity of Ship

### 4.10.3.2 Accident Frequency and number of Ships

Figure 4.56 a, and b, shows the plot of frequency and when the number of vessel. Figure 4.56a shows plot of Fa and number of ships for 3 knot and 12 knot. According to Figure 4.56, the curve for 3-12knot numbers of vessel touch high risk region at speed of 20 knot. At 1e-4, 12 ships are allowed for 12 knot. It is observed that for both generic and case study, the higher the number of ships, the higher the risk of accident occurrence is observed. The graph could be used for speed definition or limit for the channel. In Figure 4.56a, it is observed that higher number of ship result to higher risk. The regression equation for 6 trend number of ship currently being maintained at Langat is y is 2E-05e$^{-0.11x}$ R$^2$ is 0.826. Figure 4.56a shows the plot of frequency and the number of vessel is changing.

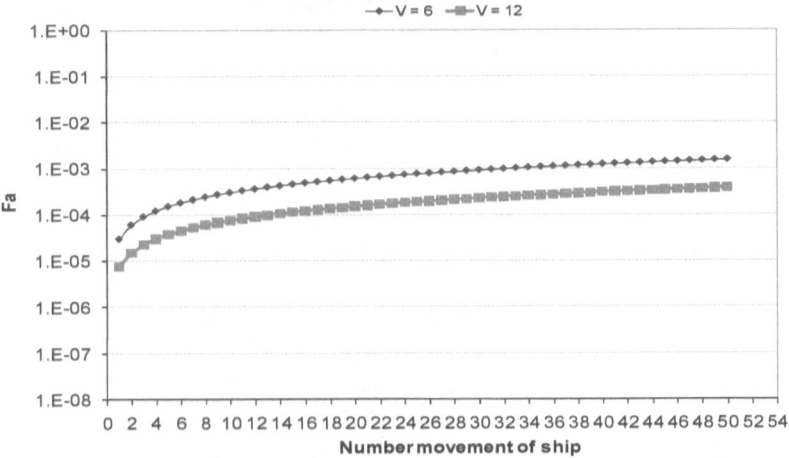

**Figure 4.56a:** Accident frequency and speed and Number of Ship

### 4.10.3.3 Accident Frequency and Velocity

Figure 4.56b shows accident frequency and speed of ship. Figure 4.56b shows changing traffic density of 6-7 ships, is acceptable and speed up to 6-8 knot, thus, this is considered high for in the Langat River waterway. But with implementation of risk control option it is acceptable. In Figure 4.56b, the risk curve movement represents ship behavior with associated changing parameter, based on that behavior; limit definition can be employed for allowable number of ships in the channel. At 12 number of ship4 knot is appropriate.

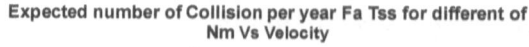

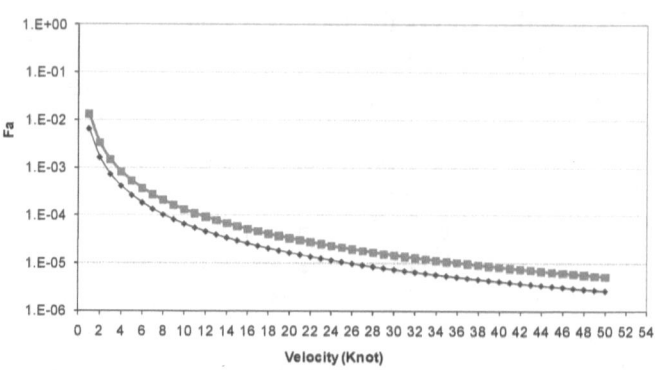

**Figure 4.56b:** Accident frequencies, speed and number of ship

### 4.10.3.4 Accident Frequency and Width/Beam

Figure 4.57 show accident frequency against channel width and beam of ship. Figure 4.57a and 4.57b show accident frequency against changing width of the channel and beam of ship, it is observed that the maximum width the channel become very risky. Width is considered three time of beam. The maximum width considered for Langat River is 64; this width is considered too small and risky for Fa at 1e-2 with maximum allowable speed of 10 knot. Thus, acceptable for accident rate of 1e-4 /year which is acceptable frequency (Fa) risk limit in maritime industry for AFA. Figure 4.57a revealed that increasing channel width to 250m above largest recommended by PIANC could allow speed of 20 knot. For current speed of 3 knot at Langat River and 12 numbers of ships, this is considered acceptable for accident 1e-4. Figure 4.57and b shows that risk is acceptable for accident per 1e-4 year, if maintenance of channel improvement plan for the River is implemented, risk lowering consideration could be discount for the current Fa value. Width of channel can change as a result of erosion. The trend for generic and Langat River revealed that as the width of the channel decrease, there is higher risk. The regression equation for the trend is represented by y is 2E-08x + 1E-05 at $R^2$ is 1. Accident frequency probability increase as wide decrease, therefore different speed should be advice to ship for such situation.

Figure 4.57b shows accident frequency and speed at changing beam of ship. Like the generic model, beam of ship is considered one third , the width of channel. At considered speed of 3 knot, and number of ships of 12 number, there is accident occurrence per 1milion years, the curve shows that the beam is acceptable at that the current speed, but could be risky, but still acceptable for accident rate per 1e-3 years for both of group of ships plying Langat River (18m for 2000t and 21 m for 5000t). Speed and number of ships can be recommended for difference beam of ship based on this trend.

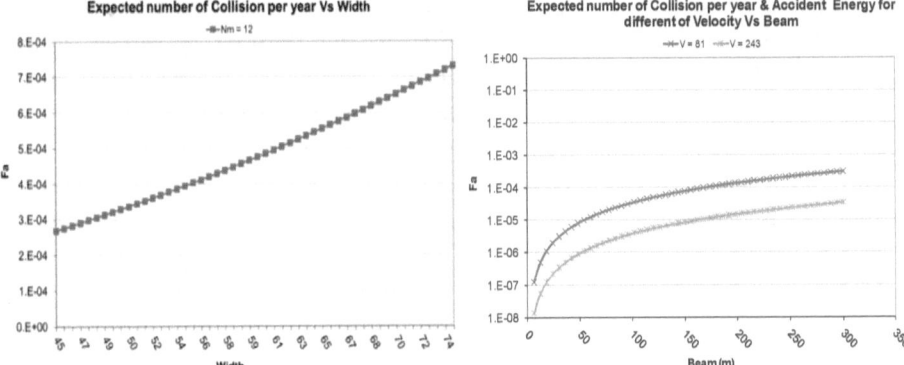

**Figure 4.57a**: Accident frequency and width of channel

**Figure 4.57b**: Accident frequency and beam of ship

### 4.10.3.5 Accident Frequency and Angle of sailing

Figure 4.58 shows accident frequency and speed at changing angle of sailing. Figure 4.58 shows AFA for angle of collision, like the generic angle Has no much effect on fa. Speed below 3 lie in high risk area, while speed up to twelve also lie within acceptable risk area.

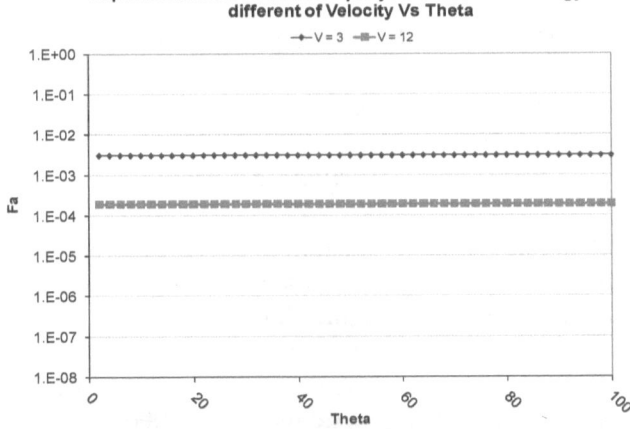

**Figure 4.58**: Accident frequency Vs Angle of sailing

### 4.10.3.6 Combined Graph of Waterways Variables

Figure 4.59 Shows combined graph of accident variables. Figure 4.59 Shows combined graph of accident variables. For accident at 1e-4, speed and number of ship and speed is critical because, they are the main two parameters that affect accident frequency in the channel. Speed of 4 knot is considered best for the channel because that point on Fa curve reflect agreement between channel parameters.

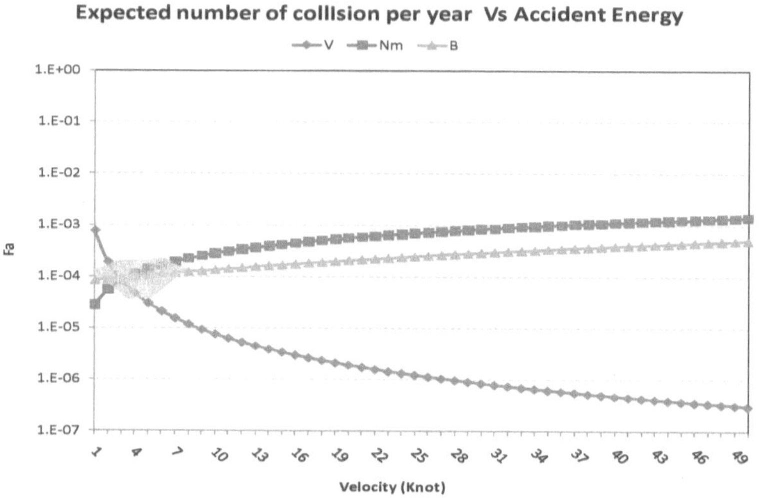

**Figure 4.59:** Combined graphs for frequency analysis with changing channel variables (V, W, Nm, B)

# 4.10.4 Result of Accident Consequence Analysis

The following shows energy, mass and velocity interaction

### 4.10.4.1 Mass of Vessel and Consequence Energy

Figure 4.60 a, b and c show variation between mass and accident energy. a: Accident consequence energy vs mass, b. Impact energy, volume of collapse and length of collapse. Figure 4.60a shows variation between mass and accident energy and velocity, for vessel of mass 5000t, there is 14 MJ of energy released, while for vessel mass of 2000t, there is 1.2 MJ of energy released at speed of 3 knot. For generic model, 20; 000 ton vessel will release unacceptable energy. Since most accident in Langat River is due to powered collision, powered collision to structure is analySed. The gragh shows that at maximum average enegry (Ca) of 36MJ, power collision energy (E1) of 400MJ, the lenght of collapse is about 11m at speed of 12 knot. Even thus this remain a totally non–linear result, the value for the case study is close to consequence analysis given for generic model. At current 3 knot speed of vessel in Langat River, insignificant length of collapse and volume of collapse is observed. The derived equation for the trend can be used for model iteration and validation of model of similar waterway profile.

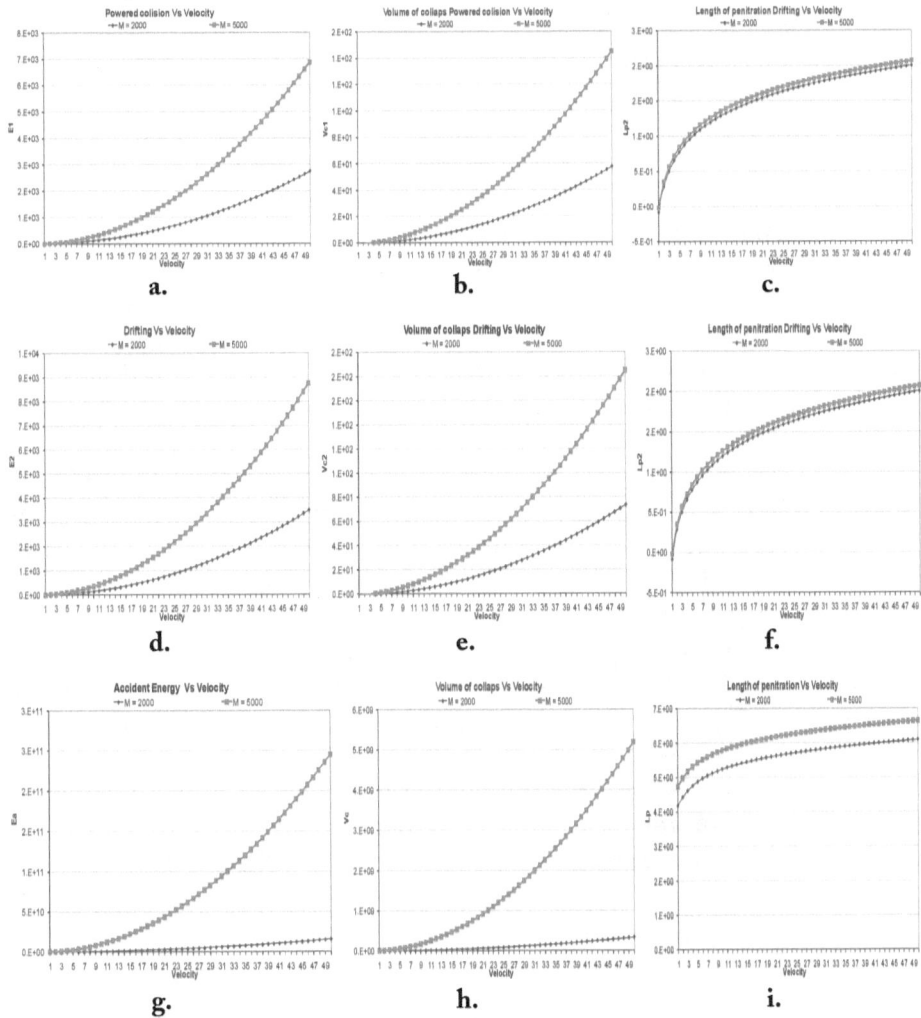

**Figure 4.60:** Variation between mass and accident energy

## 4.10.4.2 Collision Accident Impact Probability at Different Collision Situation

Figure 4.61a and b show that head on collision has the highest risk potential as the curve touch ALARP high risk area. Figure 4.61 shows the relationship between Ni, velocity, and energy interaction. Figure 4.61a shows the result of energy and speed. Figure 4.61a shows that, for overtaking collision, at 6 to12 numbers of ships that Langat River is currently operating, there is likeliness of 0 to 1 accident impact probability for overtaking collision; accident will only occur if vessel navigates at extremely low speed. Similar condition is observed for the generic model. The graph also shows that accident impact probability increase with increase number of ship. The optimum speed of ships on Langat River is 4 knot. Figure 4.61a shows that expected number of impact (Ni), accident energy at variable speed plot for head on collision, as the number of ships in the waterway increases. From 4.61b, the graph shows that there is high probability for accident occurrence between 90 and 150, where high impact can be experienced if accident happens at that angle and at speed of about 12 knot. For 2000T and 5000T barge that are currently plying Langat; speed up to 10 knot will give very high impact. It is also observed from the plot that head on collision has the highest risk potential, as the curve touch high risk area.

Figure 4.61b shows that, there is higher risk and potential rise in Ni as the number of ship in the channel increase, the graph also shows that the risk of accident of occurrence is possible when the speed is moving at much lower speed. For 6—12 numbers of ships in the channel navigating at 4 knot, there is likelihood of 0-1 accident occurrence. Also in the generic model it is observed that Ni for head and overtaking collision has higher risk than overtaking collision situation. This number is agreement with generic model where 12-24 number of ships is consider for a largest channel recommend by PIANC, about two times the size of Langat River. The graph can be use to provide corresponding speed and ship tonnage for safe ship movement in channel. Higher mass correspond to higher risk. Derived equation for each collision is shown in the graph.

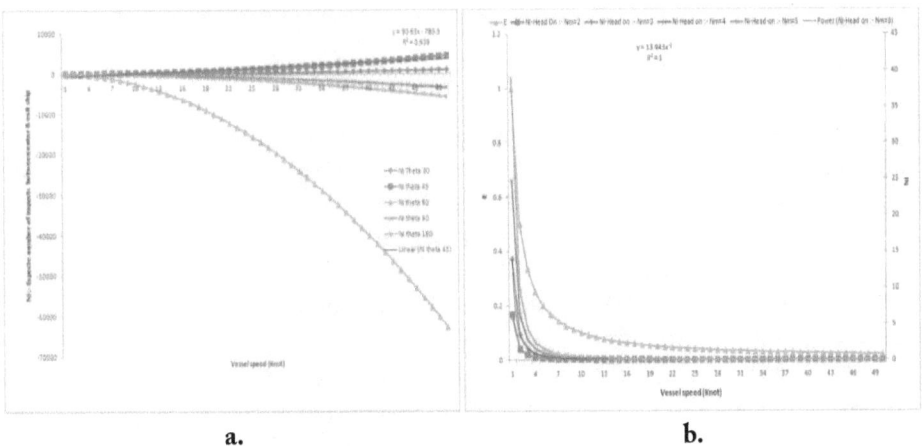

a.                                                     b.

**Figure 4.61:** a. Collision energy vs speed at different collision situation, b. Collision accident impact probability vs velocity at different collision situation

Figure 4.62 shows Number of accident impact, accident released energy and volume of collapse for different collision situation. Figure 4.62 shows the relationship between Ni, released energy and volume of collapse for different accident situation. Excess number of Ni up to 7—8 with high energy release and catastrophic volume of collapse up to 13 cubic meters can occur at the point where the energy and Ni graph meet. Figure 4.62 shows degree of risk for different accident situation. Head on collision is considered more risky as observed in the generic model. Figure 4.62 shows that accident at specific angle is harsher, whereas other accident situation is less harsh. Also, it is observed that maximum volume of collapse is at 13 cubic meters at about 400MJ released energy.

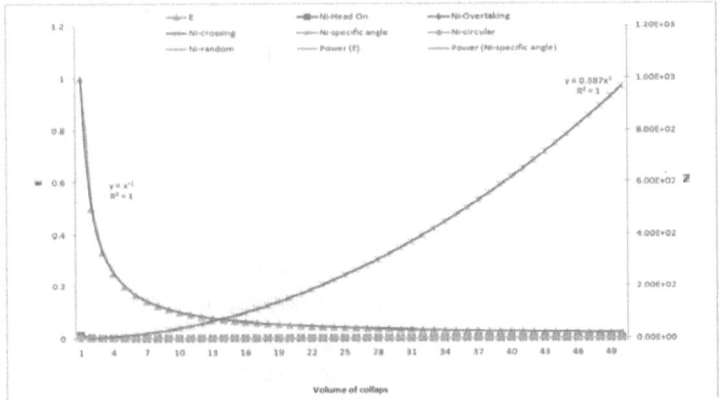

**Figure 4.62:** Number of accident impact, accident released energy and volume of collapse for different collision situation

### 4.10.4.3  Mass of Vessel and Speed at Changing Volume of Collapse, Consequence Energy for drifting collision and powered Collision

Figure 4.63a shows degree of risk for different accident situation. Head on collision is considered more risky. Figure 4.63b shows acceptable damage under different mass and energy. Figure 4.63a shows acceptable damage (Vc) under different mass and energy, acceptable damage is at at 12MJ. Figure 4.63a shows that the energy and volume of collapse relation for powered collision energy (E1), drifting collision (E2) and ship to ship collision energy (Ea). It is observed that volume of collapse of 51 cubic meter occur at 12 MJ release of energy from E1 and 27 MJ from Ea, the speed at this point is observed to be 8—12 knot, catastrophic energy will occur at speed of 38 knot, at point where E1, Ea and Vc are equal with respective value of E1 is 3000MJ, Ea is 3008MJ and Vc is 60 cubic meter. Similarly result is observed for drifting collision as shown in Figure 4.63b. The value is a bit higher for drifting colision and the gragh of drifting colision is much more steeper than the gragh of powered collision. Driftng collision E2, for exemple can be due to lost of mooring function, that can consequentially leads to vessel moving at without control to nearby structure while powered colision can be due to loss of propulsion function. Respectively subsytem level risk analysis can be perform for cause of cause function using fault three and event tre analysis to deduce much more reliability on the system.

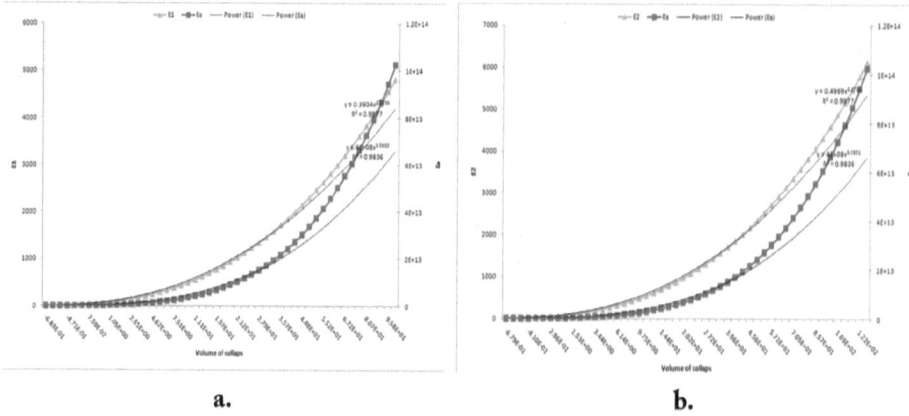

a.                                          b.

**Figure 4.63:** a. Powered collision energy **Figure 4.63b:** Drifting collision energy and volume of collapse and volume of collapse

Figure 4.64 shows interaction of vessel to vessel collision mass, energy, impact probability, volume of collapse and speed. Figure 4.64 correlations for vessel to vessel collision damage, damage are tolerable bewteen 5MJ – 20MJ.

Figure 4.64 show variation in accident consequence, mass of ship and volume of collapse, at curent speed of 3 knot the maximum mass of vessel that can give minimum volume of collapse of 0. 49 Cubic meter occur at 1500t with 9.45 MJ energy. The accepable energy for low impact event is below 50 MJ, which can lead to 6.7 volume of collapse at allowable speed of at speed of 5 knot. For vessel of of 2000t mass, there is release of 22 MJ of energy and 0.22 cubic meter of volume of collapse, vessel less than 2000ton can sail at maximum speed of 30 knot. for vessel of 5000t mass, there is release of 35 MJ of energy and 3.7 meter of volume of colapse. The derived equation for volume of colapse is y = 0.567x²—11.68x + 51.41 with corelation of 0.995 correlation. The equation derived for energy trend is y is 0.35x³ with correlation of 1, while the derived equation for mass is 500x at R² is 1. Catastrophic volume of colapse can occur at 90 cubic meter with of vessel mass up to 12300t. Also it is observed that vessel of up to 19000t could lead catastrophic relase of energy up to 19205MJ. The derived equation for the trend can be used for model iteration and validation of model of similar waterway profile.

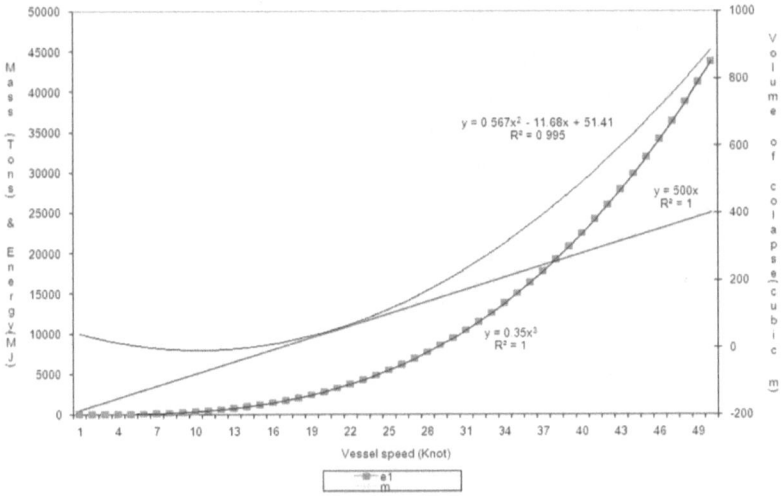

**Figure 4.64:** Vessel to vessel collision mass, energy,
volume of collapse impact probability and speed

### 4.10.4.4 Mass of Vessel and Speed at Changing Length of Collapse, Mass and Consequence Energy for drifting collision and powered Collision

Figure 4.65 shows the relationship between energy released and length of collapse. Figure 4.65 and 4.66 show the relationship between energy released and length of collapse. In Figure 4.65 acceptable damage penetration of 52m start to occur at 20MJ, unacceptable damage penetraton of 60m start to occur at 400MJ consequence energy.

Figure 4.66 shows volume and length of collapse interaction. Figure 4.66 a, b and c show interaction of volume of collapse and length of collapse, for vessel to vessel (Vc, Lp), Drifting (Vc2, Lp2) and power collision (Vc1, Lp1). High Vc and Lp is observed for drifting and powered collision, compare with vessel to vessel collision. 10 with 7m length of collapse for earlier case give 5cubic meter leading to 3m length of collapse. Figure 4.66 also shows that the length of collapse for diferent mass of veesel and respective regresion equation and correlaton. The Figure revealed that acceptable length of colapse is 7.5m at accident frequency of 10E-4. Similar comparative studies revealed the same trend on the length of penetration, where deep penetration into struck structure can start to occur at 7.18m in described in Klanac, (2005). Similar trend is observed for generic model section 4.58 in chapter Four. The consequence could further be broken down into effect for ship, human safety, oil spill and ecology.

The consequence analysis is followed by influence risk diagram that is based on ALARP principle, here the two main component (frequency x consequence) of risk is combined and plotted against each other to measure overall risk. This is followed by evaluation with risk acceptability criteria. Risk acceptability criteria established in many industries and regulations to limit the risk (Table 3.22). Risk is never acceptable, but the waterway system complexity activities imply that the risk may be acceptable due to benefits of safety, fatality, injury, individual and societal risk, environment and economy. The rationality may be debated, societal risk criteria are used by increasing number of regulators. Figure 3.22 shows risk acceptability criteria traditional rating.

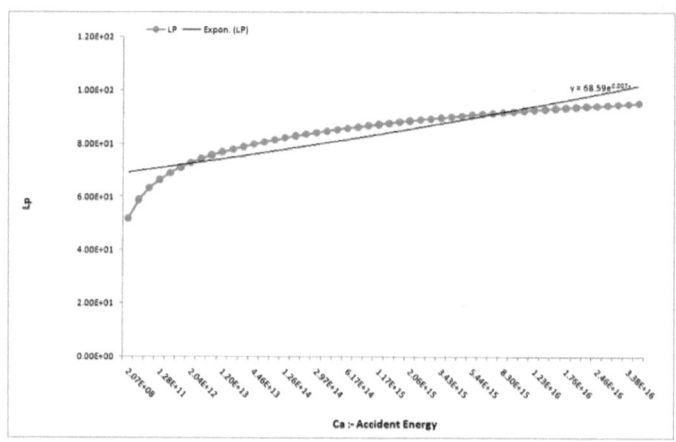

**Figure 4.65:** Length of collapse vs consequence energy

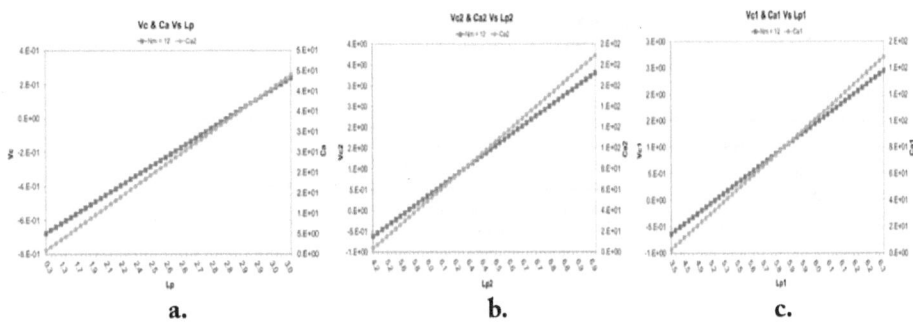

**Figure 4.66:** Volume of collapse and length of collapse

## 4.10.5  Result of ALARP

The consequence analysis is followed by influence diagram that is based on ALARP principle, here the two main component are frequency x consequence of risk is combined and plotted against each other to measure overall risk. This is followed by risk acceptability criteria. Risk acceptability criteria established in many industries and regulations to limit the risk. Risk is never acceptable, but the activity implying the risk may be acceptable due to benefits safety, fatality, injury, individual and societal risk, environment, economy. The rationality may be debated, societal risk criteria are used by increasing number of regulators.

### 4.10.5.1 ALARP Graph for Accident Frequency and Consequence Energy

Figure 4.67 a and b show the ALARP graph accident Frequency Vs Consequence. From the Figure 4.67a, ALARP comparison for Langat River risk is not appalling yet; the graph is at the lower portion of ALARP. This result on ALARP graph is within safe zone of ALARP graph, for 12 number of ships sailing inbound and outbound of the channel, at Fa acceptability of 1e-e4, less that 50MJ of energy is released, a trend similar to generic model. Figure 4.67b cross plotting shows that speed agreement of 5 knot where as for generic 7 knot agreement is achieved. Two dimension risk curve is shown in Figure 4.67b with speed agreement at about 5 knot at 1e-4. Fa and 100MJ of energy release. Figure 4.68a shows at that a speed of up 5 knot agrred speed of optimum speed for minimal accident for the waterways and energy of impact of about 137.5 MJ which is considered a moderate consequence energy. It is observed that the collision risk in Langat River is acceptable. But in the context of channel with less traffic density, the figure does not look too good. If various channel improvement plan are implemented then it is good and the level of risk will further come down.

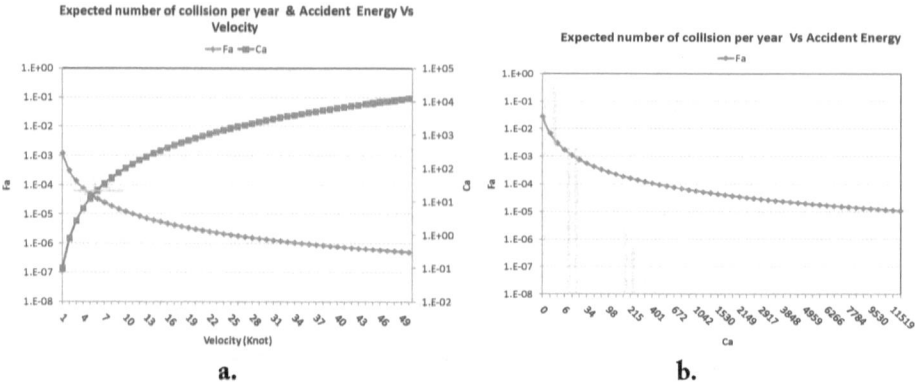

**Figure 4.67:** Accident frequency Vs Consequence

## 4.10.5.2 ALARP Graph for Accident Frequency and Consequence at changing Speed of vessel

Figure 4.68 a and b show the accident consequence energy and frequency per annum and graph. Figure 4.68b show the accident consequence energy and frequency per annum graph, the current situation on Langat River good. The situation only become risky at 0.00035 and corresponding damage consequence of 1.47E+07. Figure 4.68b is a cross plotting for accident frequency and consequence at changing speed, the trend revealed that maximum speed for Langat is 4 knot if all safety requirements are in place. The risk at that point is 1.4786E-05 x 20 MJ.

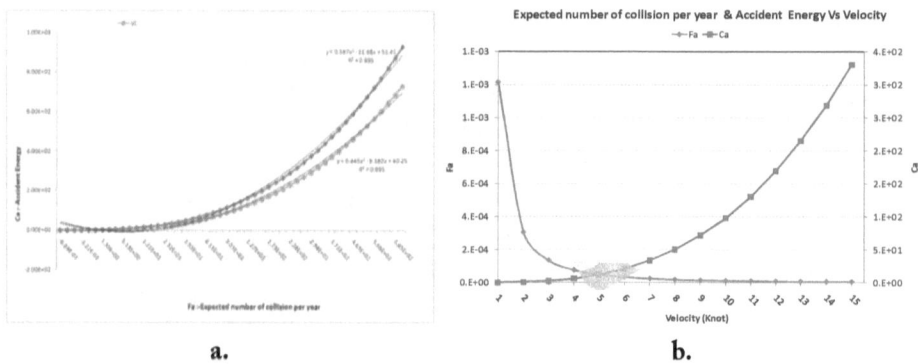

**Figure 4.68:** a. Accident frequency and accident energy,
b. Cross plot for accident frequency and consequence at changing speed

## 4.10.5.3 ALARP Graph for Accident Frequency and Consequence Volume of Collapse

Figure 4.69 shows the graph frequency and damage in term of volume of collapse graph. Figure 4.69a and b show the graph frequency and damage graph, from the graph figure, based on ALARP principle and sustainability analysis, decision can be made for waterways requirements. The ALARP graphs shows similar trend that situation at Langat is not risky. In Figure 4.69 b and c, the volume of collapse for vessel to vessel, powered and drifting collision is minimal at current speed and number of vessel at Langat at River.

a.                                          b.

c.                                          d.

**Figure 4.69:** ALARP graph of Accident frequency Vs Consequence a: Energy, b. Volume of collapse vessel to vessel collision c. volume of collapse (power collision), d: Volume of collapse (drifting collision)

### 4.10.5.4 Accident Energy and Frequency at Changing Mass

The graph for Figure 4.70 a and b shows accident energy and accident frequency at changing mass. The Figure 4.70 shows the relationship between accident frequency mass, and energy. For generic Model at accident energy of 20MJ and frequency of 10e4, mass of approximately 18, 000t is tolerable. For vessel of 2000t and 5000t currently plying Langat River, the risk is insignificant. In Figure 4.70 the damage at acceptability 1e-4 is minimal.

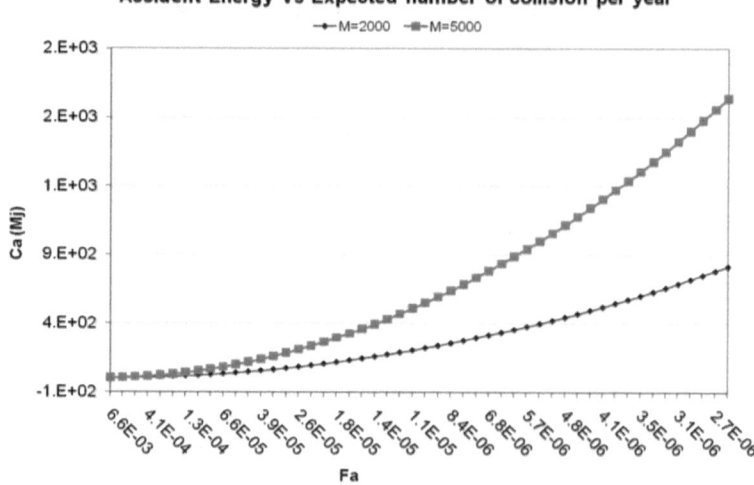

**Figure 4.70:** Accident energy, accident frequency and changing mass

### 4.10.5.5 Accident Energy and Frequency at Changing Number of Ships

Figure 4.71 shows respectively risk situation for accident frequency at changing number of ships in Langat River. Figure 4.71 shows respectively risk situation for accident frequency at changing number of ships, risk become higher with more number of ships. Figure 4.71a show that the current number of ship should be maintains on Langat River (6 ships inbound and 6 ships outbound), increase could give a sharp rise risk. Risk at Langat River become higher with more number of ships. The maximum speed that should not be exceeded is when Fa is 2.06E-04, and Ca is72E-6MJ. For the current speed of 3 knot for 5000t vessel, at Langat River, Ea value is 14MJ, and Fa is 3.8e-5. For generic 12-24 numbers of ships lies in risk acceptability region.

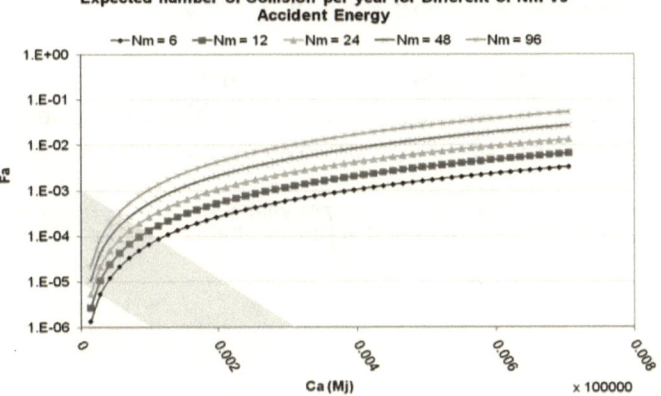

**Figure 4.71:**. Accident frequency and consequence energy at changing number of ships

## 4.10.5.6 Accident Energy and Frequency at Changing Beam and Width

Figure 4.72 shows accident frequency and consequence for width relation. Figure 4.72 shows that the current width of the channel is only good for the next 100 years; it is unacceptable for current acceptability criteria of 10e-5. With this, risk definition for width of the channel can be provided. Figure 4.72 shows that the current width of the channel (64m) is only good for the next 100,000 years. With this risk definition for width of the channel can be provided. Like generic case, width is considered 3 time beam. Figure 4.72 shows that risk become higher at large beam of ship, current beam of 18m and 21m m ship should be maintain, increase in beam of ship could lead rise in risk.

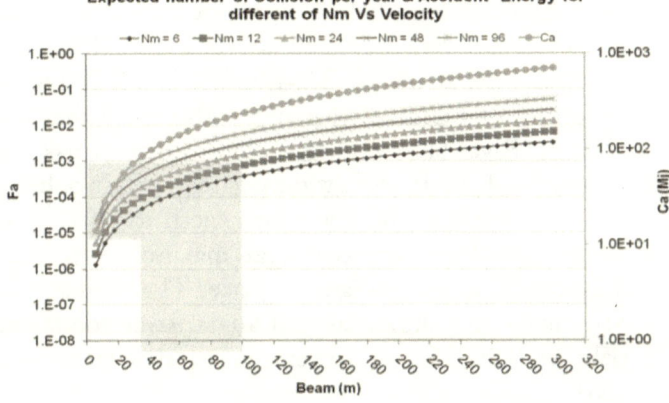

**Figure 4.72:** a, b, c and d: Accident energy vs accident frequency @
changing width of channel.

### 4.10.5.7 Accident Energy and Frequency at Changing Angle

Figure 4.73 a and b shows two dimensional plot for Fa and Ca. Figure 4.73 shows plot of risk for angle. Collision is likely to be intense at angle 50-60 for six number 6-12 number of vessel on Langat River.

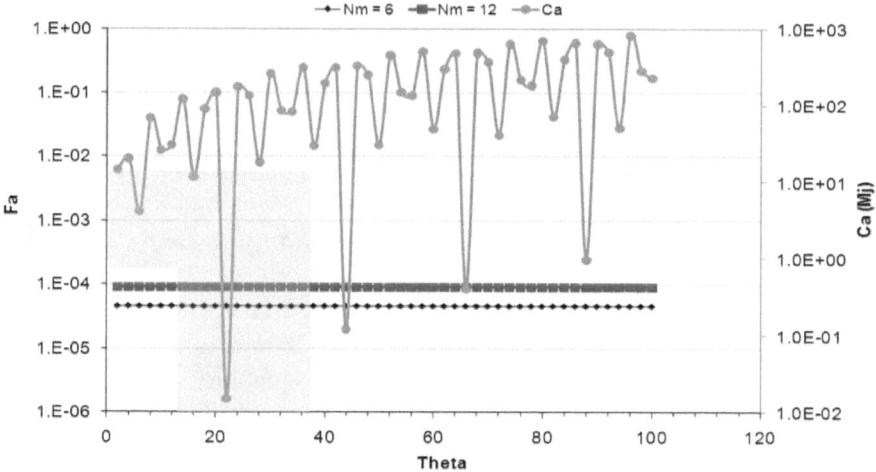

**Figure 4.73:** a. ALARP two dimensional plot for Fa and Ca.

### 4.10.5.8 Combine plot of variable for Accident Energy and Frequency

The cross plot shown in Figure 4.74 represent the combine plot of most variable of consequence analysis described above. Figure 4.74a and b shows combined plot of variables, at 64MJ, E1, E2, EA is 82MJ and when Fa is 2.8 E-6. Cross plotting of Figure 4.74 c and d also show that risk level for all variables on Langat River fall within acceptable range of risk acceptability. This represents point of limit definition against catastrophic release of energy. There is agreement that the maximum speed for Langat should stay between 3-4 knots. The trend revealed that maximum speed for Langat is 4 knot if all safety requirements are in place. The risk at that point is 3.5E-05 x 14 MJ, the current situations on Langat River good. The situation only become risky at 0.00035 and corresponding damage consequence of 1.47E+07 or 14MJ.

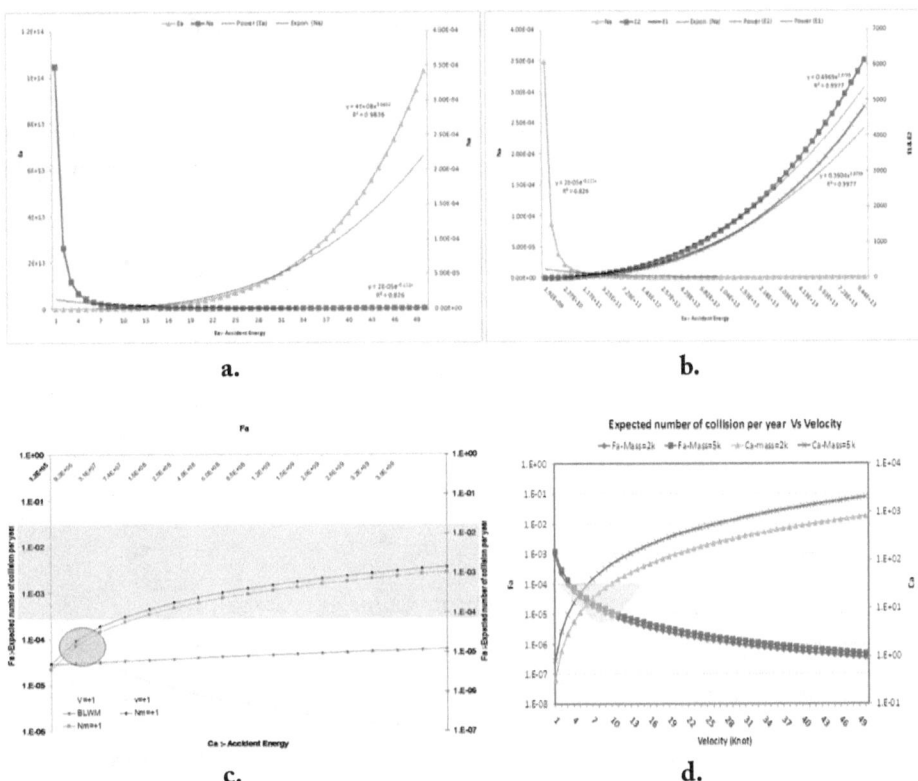

**Figure 4.74: a.** Combine plotting for accident frequency and accident consequence energy Vs consequence energy and changing beam of ship, b. ALARP graph and volume of collapse damage estimation for powered collision and drifting collision, c. Combine plot for waterway variables, d. Combine plot for 2 dimensional Fa and Ca,

## 4.10.5.9 Result of Other Consequence Quantification

The following section present result of other consequence transformation.

### 4.10.5.9.1 Volume of Collapse and Length of Collapse

Figure 4.75a, b, and c show expected volume of collapse for powered collision Vc1, drifting collision Vc2 and 1p. Figure 4.75a and b show interrelation for volume of collapse and damage estimation for powered and drifting collision. Figure 4.75a and b show expected volume of collapse for powered collision Vc1, drifting collision Vc2. The trends of the graphs are similar; however, it is observed that highest risk is attainable when with accident 1 million years for each case. Similar comparative studies revealed the same trend on the length of penetration, where deep penetration into struck structure can start to occur at 0.7m.

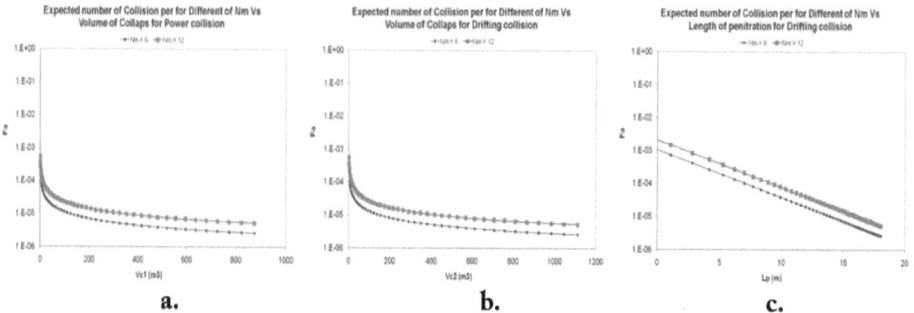

a.                b.                c.

**Figure 4.75:** Damage of volume of collapse

### 4.10.5.9.2 Fatality Analysis

Figure 4.76a and b show expected fatality in Langat River. Figure 4.76a show expected fatality in Langat River. 0 Fatality at 50MJ release of energy for current 12 numbers of ships using the channel, comparing the fatality with implementation of TSS revealed that more fatality can occur; this is due to the small size of the channel width, which means that the channel is not suitable for TSS. Figure 4.76a shows fatality risk relation, the graph shows that risk is acceptable at least no fatality (0.11), 0 Fatality at 50MJ release of energy for current 12 numbers of ships using the channel, two dimensional plots for prediction COx. Figure 4.76a shows that fatality agreement for 12 ships at risk acceptability, this signify that there is room for Langat River to accommodate more ships. Figure 4.76 c, d, e, and f show carbon dioxide release, the graph show that for the current number of ships on Langat River, the impact is acceptable, but at long run, it is unacceptable. Figure 4.76 c and d show that, for the current number of ships on Langat River, the impact for Cox releases is acceptable for a short term, but at long run, it is unacceptable. Figure 4.76c and d shows release of 5 ton per year maximum r released is acceptable for Langat River.

Cost is model in different way, translation of quantity is allowed between benefit, damage, oils spill, fatality. For this case, based on estimate on the level of damage, 1 fatality is considered high due to frequency of accident. According to current practice within IMO and selected criteria for this study, a risk control option will be regarded as cost effective if it is associated with GCAF ≤ USD 3 million or NCAF ≤ USD 3 million. Cost effective measures that can be demonstrated to have a high potential for risk reduction will consequently be recommended for implementation. ICAF represent estimation of benefit of avoiding damage or fatality and it play important role in cost benefit analysis of risk. NPV led to CURR calculation, it considers accident frequency Fa. (Number of ships per year) x Consequence Ca.x (Cost of damage per accident) before risk and after implementation of safety measure, it also consider benefit of reduced risk. NPV of the benefit for estimated risk and implemented safety measure is calculated and ratio cost of C to benefit B is compared and expected to be < 1.

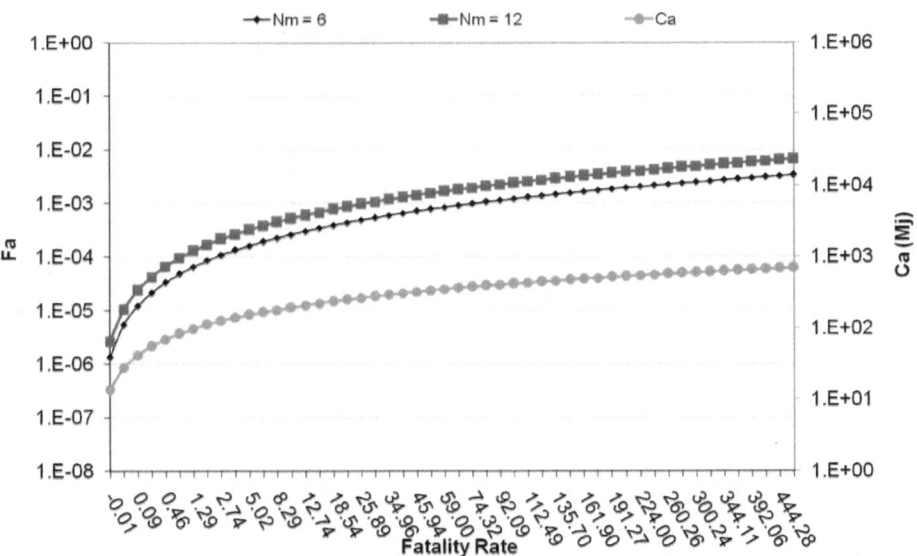

**Figure 4.76:** Fatality at Fa and Ca

### 4.10.5.9.3 Green House Gas (COx)

Figure 4.76 shows Cox release estimation.

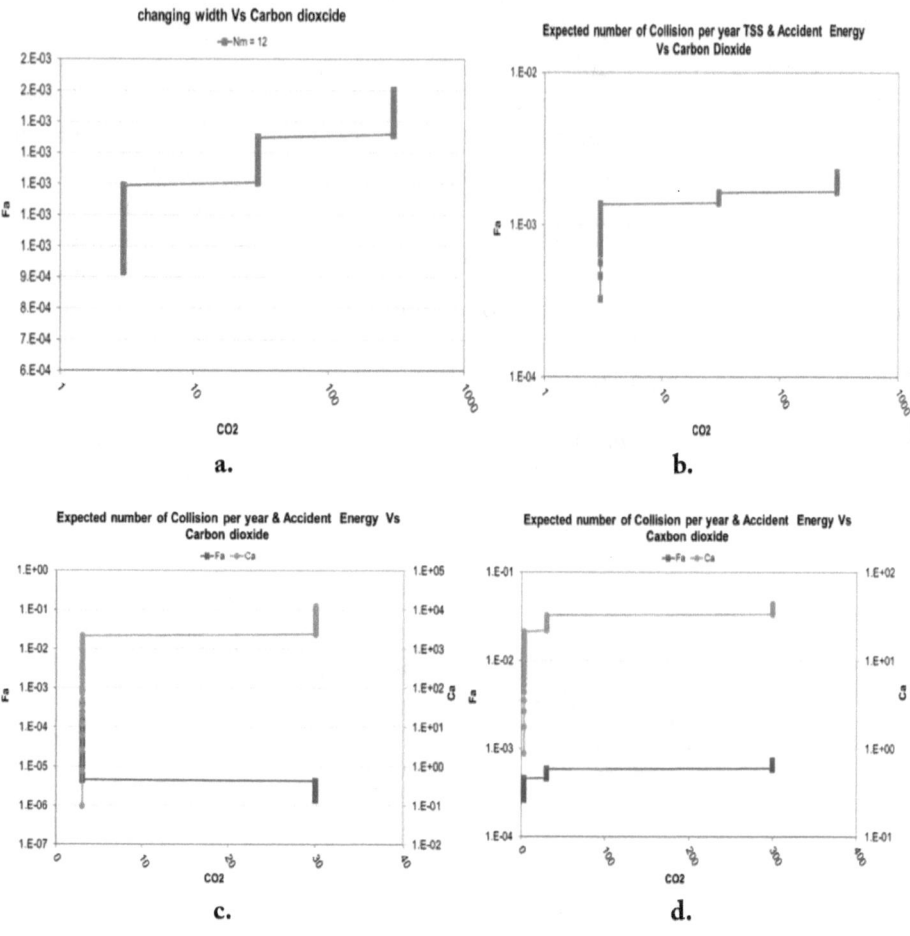

**Figure 4.76 a-d:** Consequence of COx release estimation

## 4.10.6 Cost Effectiveness Analysis

Cost work is model in different way, and translation of quantity is allowed between benefit, damage, oils spill, fatality. Figure 4.77 shows evaluation of cost of speed, the cost of speed shows the same trend with generic model. The Figure 4.77 describes the reliability cost for waterway variables. Figure 4.77a show total cost Ct, Cost of avoiding damage Co, and cost of repair Cc. Best cost stand at 1.5million at damage of 400-600MJ, a good sign for a strong model. Figure 4.77a shows that at minimum energy of 20MJ, less than 700, 000 RM gross costs will be required to avert fatality, whereas, at 400 MJ energy of impact 2.07 million RM, gross cost will be required to avert fatality. Figure 4.77 b shows evaluation of cost of speed, the cost of speed shows the same trend with generic model. Figure 4.96b shows evaluation of cost of speed. Here GCAF (1.6Million RM) for speed of 22 knot is slightly higher than NCAF (1.4 Million) for speed of 36 knot. Buying such speed will required speed for sophisticated waterway facilities. Figure 4.78 show GCAF, NCAF, ICAF and CURR relationship.

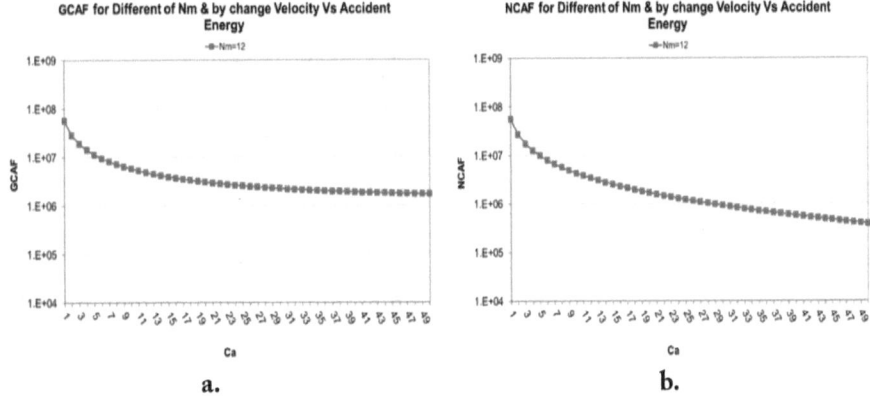

**Figure 4.77:** Cost effectiveness analysis

Figure 4.78a, b, and c and d show GCAF, NCAF, ICAF and CURR relationship. Figure 4.78 depicts the cost of losses per accident causal factors. Figure 7.78a shows that GCAF (10 Million RM) to avert current 34MJ, assumed to be 1 fatality. this is higher than NCAF, Figure 4.78 shows that 1.47 million net cost will be required to avert fatality at minimum energy of 20MJ, 2,8 million RM will be required to avert fatality at catastrophic accident energy of 453MJ. ICAF and CURR are around 1 million RM. Figure 4.78 shows evaluation of cost of speed, GCAF is about 3 million while NCAF is 2 million (Figure 4.77b). CURR is supported by ICAF is considered within acceptable cost range, below 3 million. For mass of the channel, acceptable mass is 15,000t, at the point when the cost of repair and the cost of damage are equal (50 million), the total cost 150 million (Figure 4.78c and d). At that time NCAF is about 100milion. GCAF is 80 Million and ICAF IS 50 million. This is considered very high and unacceptable. Figure 4.78c shows that 731,000 RM will be ICAF required at minimum accident energy released of 20MJ while, 2.53 million is the maximum acceptable, ICAF of released energy of 453MJ, social cost, number of injuries or fatalities, daily rate of interest. Duration of damage, sick leave in day, 6000 days is equivalent with 1 fatality is taken into consideration. Cost of damage per day depends on types and countries.

Table 4.7 depicts the cost of losses per accident causal factors. The cost of speed shows the same trend with generic model. All numbers are based on introduction of one RCO. Introduction of more than one RCO will lead to higher NCAF and GCAFs for other RCOs addressing the same risks. High GCAF and NCAF values indicate that the considered RCO is not a cost effective measure. A negative NCAF indicates that the RCO is economically beneficial, For example the costs of implementing the RCO are less than the economical benefit of implementing it. From the Figure, number of accident and loss of life are considered low. The cost of risk, can be translated to deduce benefit in different consequence quantity in term of cost, damage, oils spill and fatality.

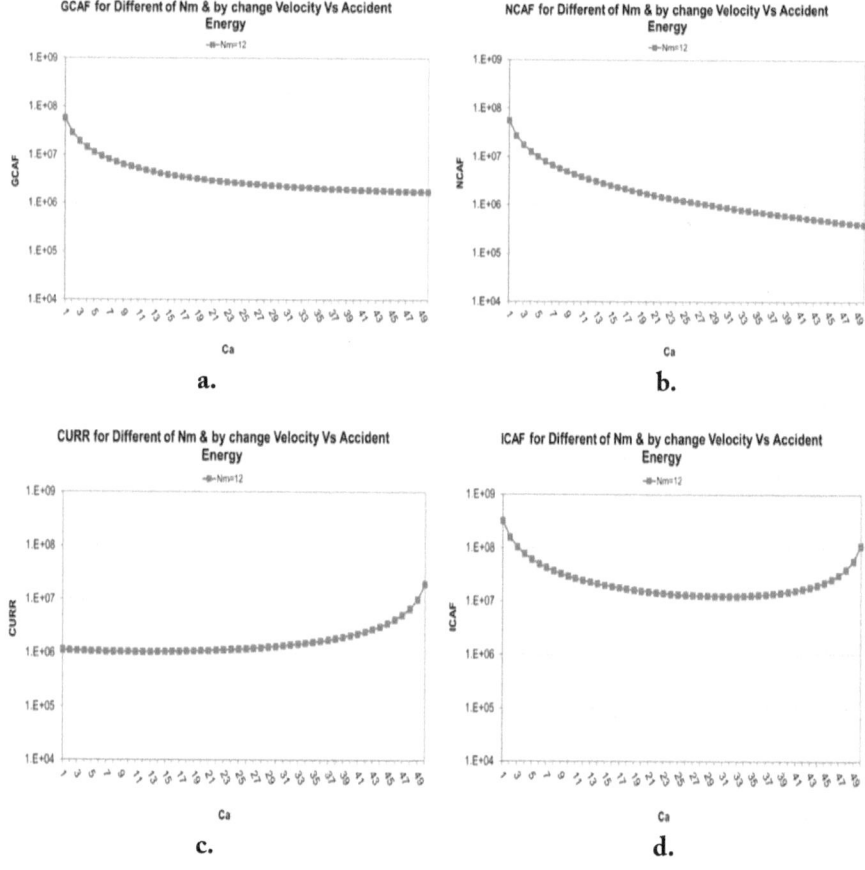

**Figure 4.78:** Cost of Averting Fatality

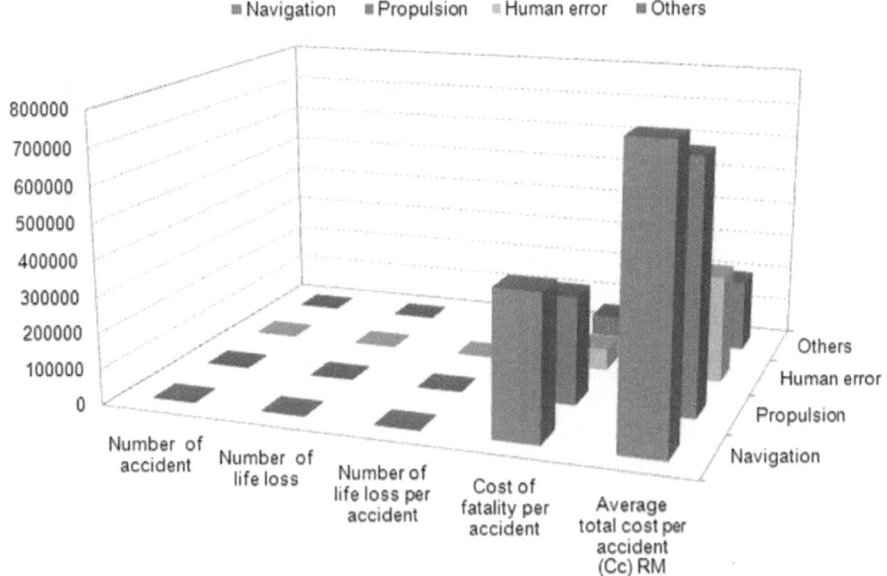

**Figure 4.79:** Cost of losses per accident causal factors

Figure 4.79 shows cost of losses per accident causal factors. Figure 4.79 show it cost much more to implement navigation and machineries failure system. Figure 4.79 depicts the cost of losses per accident causal factors. Propulsion failure carry the highest (RM 2,000,000) follow by loss of navigation function which require about RM 700, 000 and about 400, 000 will be require to fix human error problem. These costs are still acceptable as long as they are less than 3 million.

### 4.10.6.1 Cost Benefit of Damage for Optimal Energy Capacity

Figure 4.80a shows RC0 analysis for alternative option. Figure 4.80a shows RCO`s analysis for total cost of damage. Cost of unit risk reduction (CURR) of 1.47 million is required to reduce the risk for minimum release of accident energy of 20MJ and 2.85 million for catastrophic accident of 453MJ. Figure 4.80a shows RC0 analysis for alternative option. Figure 4.80b shows risk cost benefit and sustainability analysis According to recent discussion with Langat River, a decision is already made to bypass the bridge over the river. Therefore for Langat River that need to be included in analysis, but benefit could be quantify into cost. Figure 4.79 revealed losses associated with Langat River accidents. Consideration is given to cost of equipment, redesign and construction, documentation, training, inspection maintenance, auditing, regulation, reduced commercial used and operational limitation like speed and loads. Benefit could include reduced probability of fatality, injuries, serenity, negative effects on health, severity of pollution and economic losses. Table 4.8 shows summary of variable and risk level for Langat River.

The result of risk analysis carried out on IWT demonstrate that the risk is below the ALARP region, there may be no need to conduct cost effectiveness analysis, but the fact that the waterways of that magnitude is underutilized and lack of basic infrastructure and system, regulation and monitoring has contributed to historical increased in yearly accident recorded. This necessitate need to conduct RCO, CEA measure and sustainability for future installation of safety and environmental infrastructure. Figure 4.80a shows that it cost much more to implement navigation and machineries failure system. The maximum cost is indicated by the point where the total cost (Ct), the present value of loss, and NPV coincide, about RM30 million, where the cost of unit risk reduction still stand at about RM2, 000, 000.

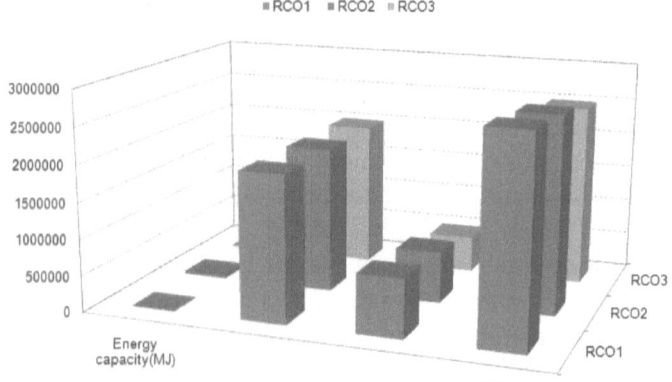

**Figure: 4.80a:** RCO`s analysis for total cost of damage

## 4.10.6.2 Cost Effectiveness Analysis of Selected RCOs

Figure 4.80b shows cross plot of the risk level and optimal cost require for the channel maintenance. Figure 4.80b shows cross plot of the risk level and optimal cost require for the channel maintenance. Figure 4.80b shows cost benefit analysis, cost agreement between the cost of damage and repair is at 1.8 million. The graph also shows that by spending more than RM50Million, high speed craft or freighter of 35 knot is able to navigate on Langat River in future. The cost is effective only at 5.00e+06RM. Under channel complexity consideration, for Langat River, poor visibility might be expected to increase the risk of groundings and collisions. It is also observed that for Langat River, speed risky above4 knot, compare with generic model where speed is still acceptable at 7knot. The increase in accident risk due to poor visibility is more consistent and more significant than the change associated with high wind.

RCO for each collision situation is clearly defined. In order to identify new RCO, generated result from the analysis of frequency and consequence and weiging of cost and benefit is. This is could be optimized with expert rating to contribute to possible beyond compliance risk prioritisation control options for IWT on Langat River. The descriptions of the major hazards and corresponding risk control options from the hazard identification and the results from the risk analysis which are summarised could also be presented to the group of experts for further validation. From the risk study, prioritized RCO that were selected for further evaluation in terms of cost effectiveness assessment are discussed below.

The risk control option proposed for the case study is similar to the generic model because the accident trend yields almost the similar result. Even thus, this research is about collision, the impact to collision is not far from contact and grounding collision scenario. Some of the measure recommended for curbing collision accident could be used and be beneficial for contact and grounding accidents. The main RCO`S are:

i. Improved navigational safety.
ii. Redundant propulsion system: two shaft lines.
iii. Required maintenance plan for critical items as well design requirement for increase double hull width, double bottom depth or increase hull strength.
iv. Human factor and reliability is quite critical in risk work, it need to be done separately.

**RCO 1: Improved Navigational Safety:** Improved navigational safety can help reduce the frequency of collision and grounding through implementation of the following:

i. Electronic Chart Display and Information System (ECDIS) integrated with track control system.
ii. Automatic Identification System (AIS) integration with radar.
iii. Improved bridge design including ergonometric.

Improvement on radar performance and navigation can be achieved through:

i. Identification of ship names for radar targets and clarification of the target's intentions.
ii. Detection of targets which are in radar shadow areas and extends radar range.
iii. Easier access of AIS information.
iv. Accounting for the ship's rate of turn to predict accurately the target's path.

CEA is useful to deduce and proposed need for new regulations based on mitigation and options selection. CEA involve quantification of cost effectiveness that provides basis for decision making about identified RCO, this include the net or gross and discounting values for risk reduction potential benefit. DNV recommendation could also be adopted as recommended for generic model.

**RCO 2: Redundant Propulsion System:** Machinery failure is one of the significant causal factors in collision accident. Reduction in risk associated with loss of propulsion could be achieve by providing the ship with redundant propulsion or steering systems. This could provide the following advantage for risk reduction:

i. The maneuverability of the ship can be maintained.
ii. A minimum speed can be maintained to keep the ship under control.
iii. Maintenance of ship with great reliability through redundant propulsion or steering system.
iv. Improved safety levels.
v. Higher vessel availability.
vi. Protection and preservation of environment through accident free channel.

**RCO 3: Human Capital Development:** The effect of training could provide better navigational safety and subsequence reduction of collision, grounding and contact events. The simulator training could be specially designed for Langat port environments, underwater topography, and particular bridge layouts on specific vessels. This would give the participants to exercises and mastering in handling challenging situations from different positions of the bridge and team requirement. Important parts in such exercises might be passage planning, situation awareness and operation during malfunction of critical technical equipment. The risk control option suggested herein goes beyond the basic training requirements defined by IMO's International Convention on Standards of Training, Certification and Watch keeping for

Seafarers (STCW), as this rule rarely apply to inland waterway transportation. Table 4.8 show summary of risk of Langat River, Table 4.9 shows summary of derived regression equations. Accident trend for validation model follow similar trend with good curve to fit. Thus, TSS for Langat River could accommodate maximum speed of 5 knot is not recommended for present situation at Langat. The risk level for implementation of TSS pose higher risk, thus if implementation is done to section of the River with required navigation safety equipments, then TSS can work for required section.

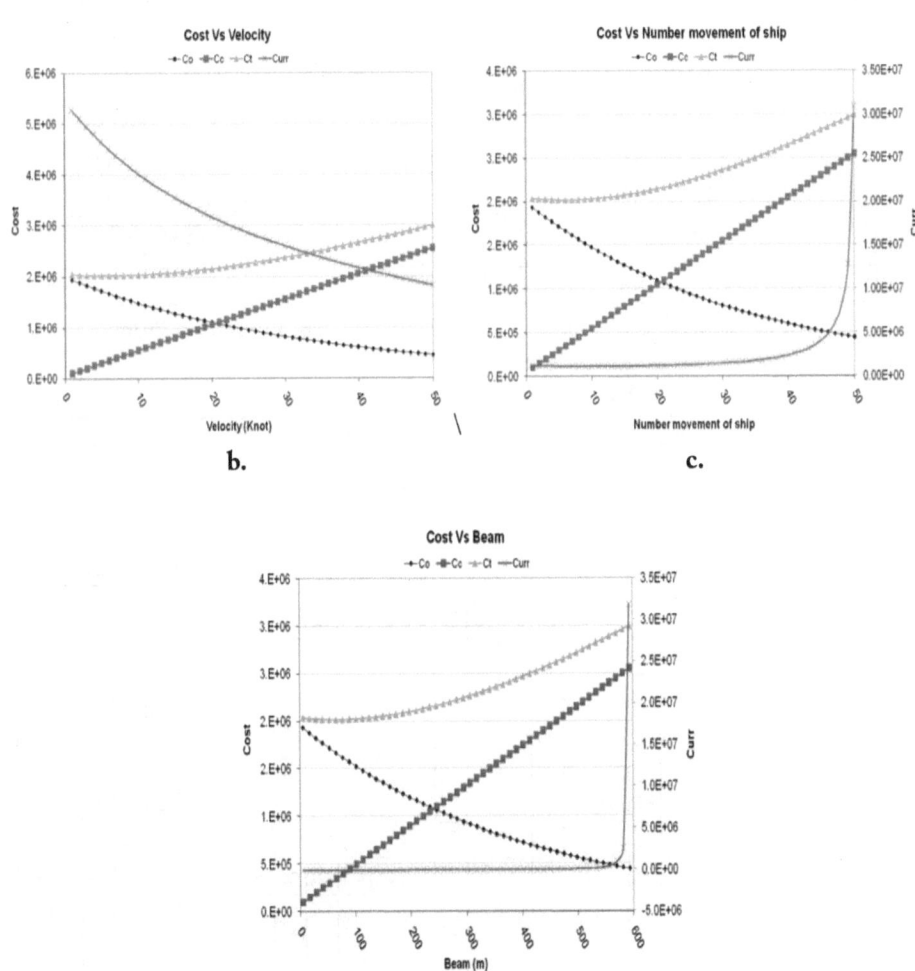

**Figure 4.80b~d**: Risk cost benefit analysis

Table 4.7 depicts losses associated with Langat River accidents

**Table 4.7:** Cost of Failure for Langat River (Personal communication (LUAS, 2008)

| Primary cause | Number of accident | Number of life loss | Number of life loss per accident | Cost of fatality per accident | Average total cost per accident |
|---|---|---|---|---|---|
| Loss of navigation control | 2 | 2 | 1 | N/A | 700 K RM |
| Loss of propulsion | 2 | 2 | 1 | N/A | 900 K RM |
| External factor | 2 | 2 | 1 | N/A | 100 K RM |
| Cause from other vessels | 2 | 2 | 1 | N/A | 100 K RM |
| Human error | 2 | 2 | 1 | N/A | 200 K RM |

**Table 4.8a:** Summary risk on Langat River

| | V | Nm | W | B | M | Fa | Ca | ICAF |
|---|---|---|---|---|---|---|---|---|
| Low | 3 | 3 | 3B | 18.3 | 2000 | 10E-4 | 10E+6 | 2 Mil |
| Medium | 3 | | 3B | 21.3 | 5000 | 10E-4 | 10E+6 | 2 Mil |
| High | 5~7 | 6 | 120 (Straight) | 40 | 10000 | 10E-3 | 400E+6 | 5 Mil |
| Critical | | | 98 (Bend) | | | | | |

Note: Criticality: V-38knot, Ca=3000mj, Pi-7-8, Vc=60, B=250m

**Table 4.8b:** Summary of Risk

| Main Variable | | | | | | | | Fa | | | | Ca | | Co | | | | | | Uncertainty | | | | | | |
|---|---|---|---|---|---|---|---|---|---|---|---|---|---|---|---|---|---|---|---|---|---|---|---|---|---|---|
| V | Nm | W | B | M | Pc | Pi | Pa | Fa | E1 | E2 | Vc | Lp | Ca | NPV | GCAF | NCAF | ICAF | CURR | NF | NaTss | CaTSS | FCRI | ETA(HRA) | FT(HRA) | ETA(P) | FTA(P) |
| 3 | 3 | 3B | 21.3, 18.3 | 5000 | 6.901 | 5.60E-05 | 2.9xCollision/pass age | 3.81E-05 | 7.5 | 9.5 | 3.1 | 71 m | 14 MJ, 1.2 MJ | | Rm 2.07 Million | Rm 2.80 Million | Rm 2.53 Million | Rm 4.10 Million | 2 | 4.97E-05 | 3.20E-03 | 0.04 | 1.12E-01 | 0.18 | 0.03248 | |
| 4 | 4 | 3B | 30 | 7000 | 0.1 | | | 1E-03 | 50 | 50 | 14 | 111 | 46.7 MJ | | | | | | | | | | | | | |
| 7 | 6 | 3B | 32 | 8000 | | 1E-02 | | | 137 | 3008 | 60 | | | | | | | | | | | | | | | |

**Table 4.9:** Summary of derived regression equations

| Plot | Variables | Equation | Correlation | Trend |
|------|-----------|----------|-------------|-------|
| **Accident impact probability** | | | | |
| Ni vs angle | Vs | $y = 90.638x - 785.53$ | $R^2 = 0.9399$ | Exponential |
| Ni Vs Pc | B | $y = 5.1138x^{-1.616}$ | $R^2 = 0.995$ | Power |
| Ni Vs Ca | W ( head on collision) | $y = 13.943x^{-2}$ | $R^2 = 1$ | Power |
| Ni Vs Fa | W (overtaking collision | $y = 1.0541x^{-2}$ | $R^2 = 1$ | Power |
| Ni Vs Fa | W( specific angle) | $y = 23.011x^2 - 476.58x + 2125.1$ | $R^2 = 0.9958$ | Power |
| Ni vs Fa | Nm (random collision situation) | $y = 0.0008x^{-2}$ | $R^2 = 1$ | Power |
| Ni vs Fa | Nm (circular situation) | $y = 4.6937x^{-2}$ | $R^2 = 1$ | Power |
| **Frequency model** | | | | |
| Fa | @Nm changing Speed | $y = 2E\text{-}05e^{-0.11x}$ | $R^2 = 0.826$ | Exponential |
| Fa | @V | $y = 2E\text{-}05e^{-0.11x}R^2 = 0.826$ | $R^2 = 1$ | Square |
| Fa | W | $y = 2E\text{-}08x + 1E\text{-}05$ | $R^2 = 1$ | Square |
| Na | B | $y = 9E\text{-}07x + 0.000$ | $R^2 = 0.999$ | Linear |
| **ALARP Risk curve** | | | | |
| Fa VCa | V | $y = 0.0003x^{-2}$ | $R^2 = 1$ | Power |
| Fa VCa | B | $y = 0.0001x^{-1.616}$ | $R^2 = 0.995$ | Power |
| Fa VCa | B =18.3, M=2000 | $y = 0.0003x^{-2}$ | $R^2 = 1$ | Power |
| Fa VCa | B =21.3, M=2000 | $y = 0.0004x^{-2}$ | $R^2 = 1$ | Power |
| Fa VCa | Nm | $y = 0.0014x^{-2}$ | $R^2 = 1R^2 = 0.826$ | Power |
| | Nm=3 | $y = 2E\text{-}05e^{-0.111x}$ | | Exponential |
| Fa VCa | W=30-2480, M=5000 | $y = 0.0003x^{-2}$ | $R^2 = 1$ | Power |
| Fa VCa | W=46 | $y = 0.0003x^{-2}$ | $R^2 = 1$ | Power |
| **Damage** | | | | |
| Fa Vs Lp | Speed | $y = 0.0003x^{-2}$ | $R^2 = 1$ | Power |
| Lp Vs Ca | Lp | $y = 11.219\ln(x) + 51.188$ | $R^2 = 0.9997$ | Logarithm |
| Lp vs Fa | M | $y = 5.1793\ln(x) + 63.748R^2 = 0.9988$ | | Logarithm |
| Fa Vs Vc1 | Vs | $y = 0.0003x^{-2}$ | $R^2 = 1$ | Power |

| Fa Vs Vc2 | Vs | $y = 0.0003x^{-2}$ | $R^2 = 1$ | Power |
|---|---|---|---|---|
| Ca vs Fa | Vc1(Drifting collision) | $y = 0.4457x^2 - 9.1828x + 40.251$ | $R^2 = 0.9958$ | Polynomial |
| Ca vs Fa | Vc2(Powered collision) | $y = 0.5673x^2 - 11.687x + 51.418$ | $R^2 = 0.9958$ | Polynomial |
| **Criticality** | | | | |
| FMECA vs Ca | Vs | $y = 0.0192x^{-3}$ | $R^2 = 1$ | Power |
| FCRI | Vs | $y = 0.0096x^{-1}$ | $R^2 = 1$ | Power |
| **C0st benefit and sustainability** | | | | |
| CURR | Ca | $y = 3E+07\ln(x) - 3E+07$ | $R^2 = 0.896$ | Logarithm |
| NCAF | Ca | $y = 1E+07\ln(x) - 2E+07$ | $R^2 = 0.896$ | Logarithm |
| ICAFT | Ca | $y = 2E+06x + 6E+06$ | $R^2 = 0.984$ | Logarithm |
| GCAF | Ca | $y = 2E+07\ln(x) - 3E+07 R^2 = 0.832$ | $R^2 = 0.832$ | Logarithm |

# 4.11 Channel Complexity and Uncertainty Analysis

The preceding result present result of intermediary and uncertainty analysis. The principal aim is to determine total risk through accident and frequency analysis, the result of which is presented under ALARP section. Other intermediate results are presented under channel complexity and uncertainty analysis. Collision risk contributing factor is now composed by using Relex FMEA. The Relex FMEA analysis for propulsion failure, FMEA analysis is generated. RPN calculation of propulsion failure is the high.. Table 3.10 and Table 3.11 shows consequence risk acceptability. Also causes from external sources like small craft, under which there are cause of cause, cause from other uncertainty including human error may attract separate analysis.

Channel complexity that addressed in the risk and reliability work are visibility weather, squat, bridge, river bent and human reliability. Figure 3.32b shows channel complexity for Langat. Poor visibility and the number of bend may increase the risk of collisions. Barge movement creates very low wave height and thus will have insignificant impact on river bank erosion and generation of squat event. Speed limit can be imposed by authorities for wave height and loading complexity. Human reliability analysis is also important to be incorporated in the channel complexity risk work; this can be done using questionnaire analysis or the technique of human error rate prediction. For total risk work the following analysis could perform separately as part of subsystem risk level analysis: Power transmission, navigation, vessel motion and human reliability. These sub—causal factor can be reinvestigated by using frequency calculation through Fault Tree Analysis (FTA) modeling involve top down differentiation of event to branches of member that cause them or participated in the causal chain action and reaction. While consequence calculation can be done by using Event Tree Analysis (ETA), where probability is assigned to causal factor leading to certain event in the event tree structure. Fault Tree Analysis (FTA) for Propulsion Failure involve top down differentiation of event to branches of member that cause them or participated in the causal chain action and reaction. It is a systematic inclusion of all components' involved in particular events or failure says propulsion failure. Cumulative of FTA analyses leads to scenario generation by using minimum cut set of accident cause. Highly probable cut set will satisfy the condition that (Refer to Figure 4.27 and Table 4.11 and 4.12), the Frequency of a minimal cut set is equivalent to Total accident frequency) divided by 4.

## 4.11.1 Propulsion Failure

Figure 4.81a shows deduced collision risk contributing factors. Figure 4.81a,b,c, Table 4.11-4.13 show FTA and ETA analysis for propulsion failure. It shows probability of 0.033 , whereby loss of power contribute 0.01, gear failure contribute 0.015 and loss of main engine contribute 0.002 and clutch failure result have probability of 0.074 and main engine failure contribute 0.005, where the potential failure modes for the process of collision, the effects and the causes for each of failure modes are identified and described. Similarly FMEA is carried out to examine Human reliability. The result revealed high risk which is shown in Table 4.14 and Figure 4.83. ETA analysis also shows that collision leading o heavy damage much damage as a result of near coast obstacle carry high probability of 0.18 while collision with minor damage come from situation like anchor failure (p=0.02). Other observed probability is p=0.05 as result of heavy weather and p=0.03 as result of stall machineries.

The input to defines the limit of probability of event. Logic combination of gates couple with series and parallel system arrangement of the system is used to determine the probability. In this case the probability for propulsion failure modeled from RELEX software is 0.0325 for main cut set. The losses come from loss of propulsion power (0.009), loss of gear failure (0.015), starboard main engine failure (0.002), and Clutch failure (0.007), loss of starboard main engine (0.06), loss of power transmission (0.09x0.025), loss of starboard main engine (0.002 x 0.007), loss of port main engine (0.0.6 x 0.07). The probability for other contributing factor to this FTA is shown in Figure 4.81c.

Event Tree Analysis (ETA) for propulsion failure involve consequence calculation through event tree analysis (ETA), ETA is inductive binary method that describes the relation between initiating event and the events that describe the possible consequence. Here, each level in the chain of event leading to consequence consists of two mutually exclusive dichotomies whose probabilities are success or failure is estimated (See Figure 4.82). Event tree probability (for collision heavy damage) Pe = 0.4x 0.5x 0.9=0.18. Events probability can be calculated and each effect deal with according to potential of occurrence leading to specific impact for major accident. The Event three probabilities are calculated by multiplied successive probability of the tree. The event tree probability leading to major damage is shown in Table 4.12.

Fault Tree Analysis of HRA for Collision Top Event involve analysis of collision casualties arise due to interaction of human behavior with functions, such as function of navigation equipment and inadequate of situation awareness. The parameters for occurrence or extend of casualty are performance of machinery or equipment, level of training of crew, and poor management of team. Better training and procedures can help to promote better communications and coordination on and between vessels. Figure 4.83 and 4.84 show result of FTA modeled from RELEX software for HRA. Poor general knowledge, experience and practical is concerned with the poor understanding of mariners of how the automation works or under what conditions it was designed to work effectively. Consequently, sometimes the mariners can commit errors in using that equipment due to insufficient training.

The effectiveness of this important safety feature would be improved through the development of a performance standard. From Figure 4.83a it shows that with the severity ranked at 7, the occurrence ranked at 6 and with the highest RPN, 126. It is considered in hazardous situation where necessary actions should be taken. From the Figure 4.83b, the graph of FMEA risk level showed that with the severity ranked at 6 and occurrence ranked at 2, this failure mode "inexperience, lack of knowledge and training" has the lowest RPN, 72 compared to others potential failure modes. Nonetheless, actions such as better and adequate training, standardized equipment design and an effective method of assigning crew to ships can be quite useful to overcome this problem (Table 4.14).

Poor equipment design represents a causal factor in major marine casualties. A proper consideration by equipment designers to factors such as how a given piece of equipment will support the mariner's tasks and how it will integrate into the entire equipment "suite" used by the mariner can be a very helpful step. Poor inspection and maintenance is another important issue because poor maintenance of navigational equipments can lead to dangerous work environments and lack of backup systems that need to carry out emergency repairs. Thus, proper use shall be made of radar equipment fitted and operational, including long-range scanning to obtain early warning of collision and radar plotting or equivalent systematic observation of detected objects. The causality derived by the fault tree is consistent with the collision definition developed by MAIB. The probability of collision is calculated by using Boolean algebra working from bottom to top.

A number of recent investigations undertaken by the MAIB have shown that ships' general alarms are frequently not sounded during or after an accident. Technological improvements may actually lead to increase in overall accident risk due to a variety of human factors, including inefficiency of navigational equipment where failure caused by faulty equipment installation, when operating under normal conditions. With the IMO requirements, it is believes that watch alarms significantly enhance navigational safety. Encouragingly, an increasing number of ships are being fitted with such watch alarm systems which providing them with a valuable safeguard against watch keepers falling asleep. Table 4.14 shows FTA for HRA of collision event, Figure 4.84 shows FTA Gates for the Top Event of HRA analysis of collision accident. Table 4.15 shows Fault tree calculation for collision TE.

FTA Numerical Calculation of ETA (Figure 4.84) involve the identifier, capital letters in rectangles denote intermediate and basic fault events associated with the process respectively. Each of the capital letters is defined FTA gates. The probabilities of occurrence of independent events E1, E2, E3, E4, E5, E6, E7 and E8 are 0.008, 0.01, 0.04, 0.01, 0.04, 0.009, 0.009 and 0.009 respectively. The probability of occurrence of the top event for Collision, is calculated The probability of occurrence of AND gate, event G5 is

$$P(G5) = (0.009)(0.009) = 8.1 \times 10e\text{-}5,$$

the probability of occurrence of event G4 is:

$$P(G4) = 1 - [1 - P(G5)][1 - P(E6)]$$

Where: $P(E6)$ is the probability of occurrence of Event 6. By substituting the given values of $P(G5)$, and $P(E6)$ into Equation for

$$P(G4) = 1 - [1 - 8.1 \times 10^{-5}][1 - 0.009] = 0.00908$$

The probability of occurrence of event G2 is

$$P(G2) = 1 - [1 - P(E1)][1 - P(G4)][1 - P(E2)]$$

Where: $P(E1)$ is the probability of occurrence of event1, $P(E2)$ is the probability of occurrence of Event 2. From the above calculated and given values of $P(E1)$ and $P(E2)$ respectively into Equation

$$P(G2) = 1 - [1 - 0.008][1—0.00908][1 - 0.01] = 0.0268$$

The probability of occurrence of event G3 is:

$$P(G3) = 1 - [1—P(E3)][1 - P(E4)][1 - P(E5)],$$

Where: $P(E3)$ is the probability of occurrence of Event 3, $P(E4)$ is the probability of occurrence of event 4, $P(E5)$ is the probability of occurrence of event5. From the given values of $P(E3)$, $P(E4)$ and $P(E5)$, is

$$P(G3) = 1 - [1 - 0.04][1 - 0.01][1 - 0.04] = 0.0876$$

For the above two calculated values of $P(G2)$ and $P(G3)$, respectively, the probability of occurrence of the top event: Collision $P(G1)$ is

$$P(G1) = 1 - [1 - P(G2)][1 - P(G3)] = 1 - [1 - 0.0268][1 - 0.0876] = 0.1121,$$

The probability of occurrence of the top event for collision is 0.1121.

Human reliability analysis is also important to be incorporated in the channel; complexity risk work, this can be done using questionnaire analysis or the technique of human error rate prediction THERP probabilistic relation. Technique for Human Error Rate Prediction (THERP) is one of the best known and the most frequently applied technique for human performance reliability prediction. THERP is use to predict human error rates in association with factors such as equipment reliability, procedures and other factors. THERP can be calculated from:

$$Q_i = 1 - (1 - F_i P_i)^{n_i}$$

Where: $F_i$ is the probability that assigned a value, $P_i$ is the probability that an error will occur, $F_i P_i$ is the joint probability that an error will occur and will lead to system failure, $1 - F_i P_i$ is the probability that an operation will be performed that does not lead to a system failure, and $n_i$ is the number of independent operations. From the equation above, the probability of the three successful pivotal events for example radar detection, VHF and watch alarm system will work together is determine by:

$$Q_1 = 1 - [1 - (0.1)(0.5)]) = 0.005$$

In case the radar detection, VHF and watch alarm system do not work at the same time, the probability that a class of errors will lead to system failure is

$$Q_2 = 1 - [1 - (0.1)(0.5)(0.2)(0.4)] = 0.004.$$

Event Tree Analysis (ETA) deal with breaking down the pivotal event scenario definitions and presents this information in a tree structure that classifies scenarios according to each consequence. The headings of the event tree are the initiating event (IE), the pivotal events (PE) and the end states. The tree structure shows the possible scenarios for radar detection works, VHF works, and watch alarm system works, in terms of the occurrence and nonoccurrence of the pivotal events, computing the success or fail probability for each contributing PE that the PE states must always sum to 1.0. Since, $1 = P_S + P_F$, the probability of success is derived from the probability of failure calculation. The probability for a particular outcome is computed by multiplying the event probabilities in the path. In this case, there are three contributing PEs that generate five possible different outcomes, while each with a different probability. From Figure 4.85, when there is all success for the three pivotal events, it leads to only limited damages for the vessel and the probability is 0.05. On the other hand, if the three pivotal events are in failure modes, the consequence is death or serious injury and extensive damage will occur with the probability of 0.004.

## Table 4.10: Failure mode effect Analysis

| Identifier item | Name function | Item description | Identifier mode | Failure mode | Local effect | Next effect | End effect | Cause of failure | Severity | Occurrence | Detection | RPN |
|---|---|---|---|---|---|---|---|---|---|---|---|---|
| Failure to control Vessel motion | Vessel main engine couple with controllable pitch propeller | Propeller to transform energy, transmitted through the propeller shaft into a pressure blade, which then acceleration and maintain the speed of vessel | Mode A Propulsion | Stop/ reduce function | Gear overload | Hear Effect | Reduced speed | No fuel feed, crankshaft failure, piston hot | Catastrophic | 1 | 2 | 9 |
| | | | Mode B Gear, RPM, Transmit power | Power transmittal problem | Main engine overload | Propulsion failure | Loss of maneuvering | Broken Cog | Critical | 3 | 4 | 7 |
| | | | Mode C Shaft, line transmit power | - | - | - | - | Broken Shaft | Marginal | 3 | 6 | 4 |
| | | | Mode D CPP, transmit power | Reduce function | - | Reduced effect | Reduce speed | Broken blade | Minor | 7 | 3 | 2 |

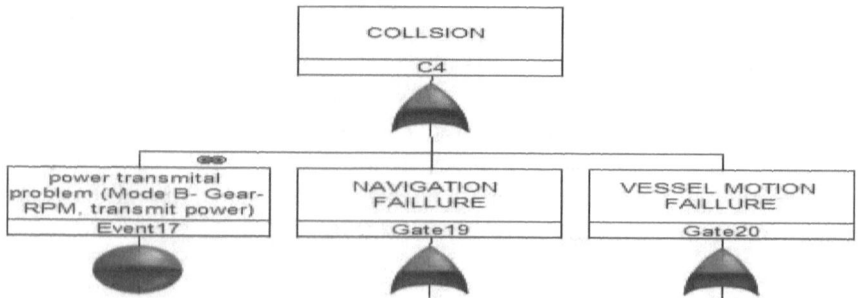

**Figure 4.81a** Collision risk contributing factors (Modeled from RELEX software)

Figure 4.81b show FTA for propulsion failure causal factors

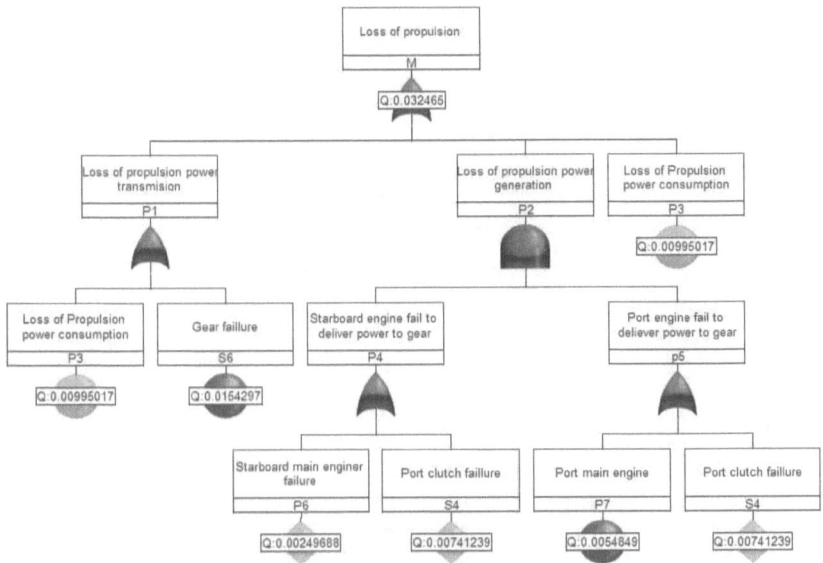

**Figure 4.81b**: Fault three analyses for propulsion failure

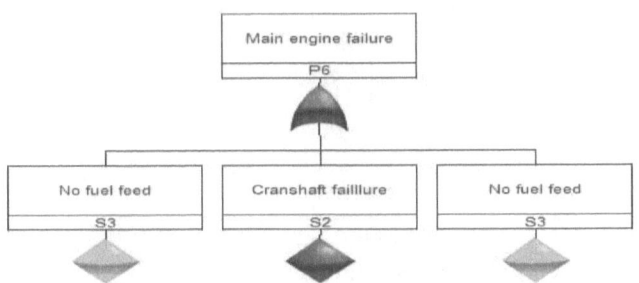

**Figure 4.81c**: Fault three analyses for sub contribution factor to propulsion failure

**Table 4.11:** FaultTree Analysis

| Report data | | | |
|---|---|---|---|
| **Name** | **Identifier** | **Reference Designator** | **Description** |
| Collision | System1 | C4 | Accident category |
| Navigation failure | System1 | Gate 19 | Main cause 1 |
| Vessel motion failure | System1 | Gate 20 | Main cause 2 |
| Propulsion failure | System2 | M | Accident category |
| Heavy/hydrodynamic weather | System3 | Event 17 | Sub causes |

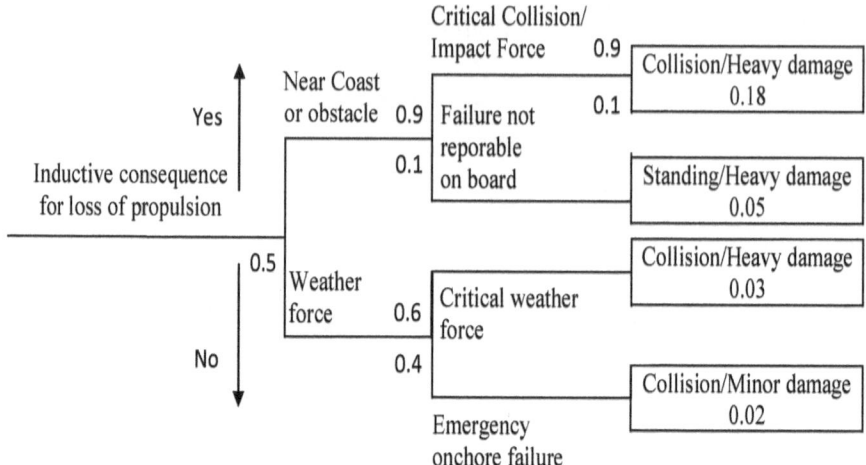

**Figure 4.82**: Event tree analyses for propulsion failure consequence

**Table 4.12**: Event Tree Analysis

| Identifier | Type | Comment | Logical Cond. | Input type | FR type | Exposure time (%) | Dormancy Factor | FR (%) | Input | Input 2 | Units |
|---|---|---|---|---|---|---|---|---|---|---|---|
| Gate 17 | Basic event | Power transmittal problem (Model-Gear-RPM, Transmit power) | | Fr/MTBF | 100 | 0 | 100 | 6.6 | | | |
| Gate 18 | Orgate | Collision | | Constant Probability | 100 | 0 | 100 | 0 | | | |
| Gate 19 | Orgate | Navigation normal failure | | Constant Probability | 100 | 0 | 100 | 0 | | | |
| Gate 20 | Orgate | Vessel normal motion failure | Normal | Constant Probability | 100 | 0 | 100 | 0 | | | |

Table 4.13: Consequence of risk acceptability

| Quantification | Serenity | Occurrence | Detection | RPN= SOD |
|---|---|---|---|---|
| current failure that can result to death failure, performance of mission | Catastrophic (10) | 2 | 3 | 60 |
| failure leading to degradation beyond accountable limit and causing hazard | Critical (7) | 5 | 5 | 175 |
| controllable failure leading to degradation beyond acceptable limit | Major ( 5) | 7 | 7 | 175 |
| nuisance failure that do not degrade system overall performance beyond acceptable limit | Minor (2) | 10 | 9 | 90 |

## 4.11.2 Human Reliability Analysis

Figure 4.83 shows graph of Failure Mode Effect Analysis (FMEA) risk level. Table 4.14 shows FMEA analysis

Table 4.14: Failure mode effect Analysis for HRA

| Process Function / Requirements | Potential Failure Mode | Potential Effect(s) of Failure | S e v | Potential Mechanism of Failure | O c c u r | Current Design Controls Prevention | Current Design Controls Detection | D e t e c | R. P. N. | Recommended Actions | Responsibility & Target Completion Date | Actions Taken | S e v | O c c | D e t | R. P. N. |
|---|---|---|---|---|---|---|---|---|---|---|---|---|---|---|---|---|
| Collision / Lack of situational awareness | Inexperience, lack of knowledge and training | Flaw decision making | 6 | Limitation of experience, knowledge and practical | 2 | | | 6 | 72 | Appropriate practical and training | | | 6 | 1 | 6 | 36 |
| | Lack of brigde team management | | 7 | Poor communication | 3 | | | 5 | 105 | Cooperation and team management is concerned | | | 7 | 2 | 5 | 70 |
| | | | 7 | Lack of briefing and preparation | 3 | | | 5 | 105 | | | | 7 | 2 | 5 | 70 |
| | Sole watchkeeper | Unaware | 8 | Insufficient member | 5 | | | 3 | 120 | Adequate personnel | | | 8 | 4 | 3 | 96 |
| Defective navigational equipment | Defective radar | Ineffective detection | 6 | Malfunction | 6 | | | 3 | 108 | Periodically inspections and maintenance | | | 6 | 5 | 3 | 90 |
| | Inefficiency functioning in VHF | | 6 | Lack of training and knowledge | 5 | | | 3 | 90 | Practical carry out frequently | | | 6 | 4 | 3 | 72 |
| | Unfitting of watch alarm systems | Unaware | 7 | Unreliable installations | 6 | | | 3 | 126 | Regularly manual checks | | | 7 | 2 | 7 | 98 |

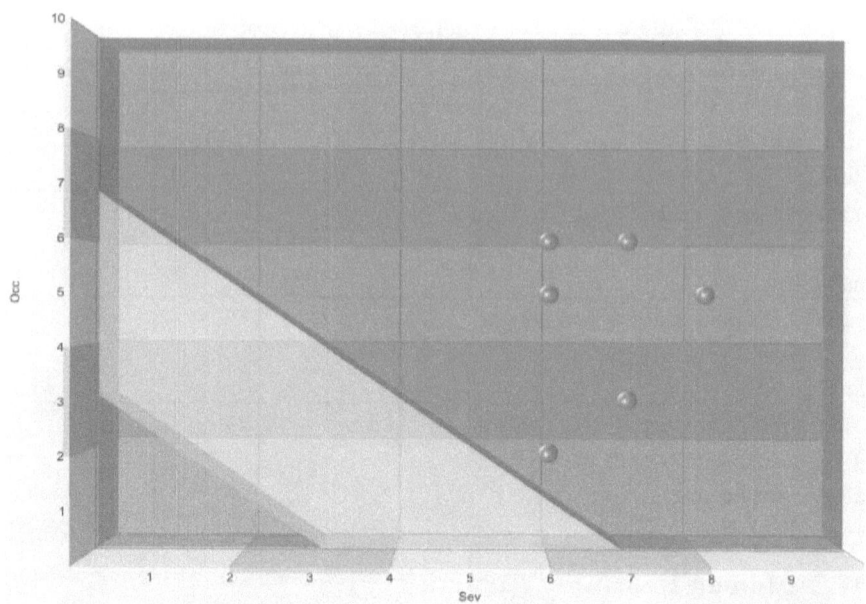

**Figure 4.83a:** FMEA ALARP measurement for human reliability risk

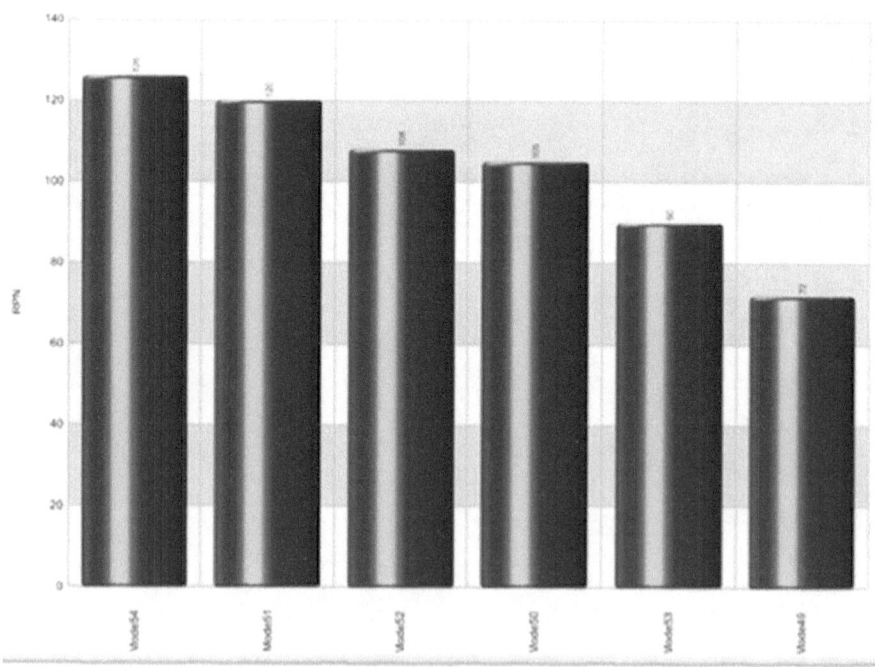

**4.83b:** FMEA risk level for HRA

Figure 4.84 shows FTA Gates for the Top Event of HRA analysis of collision accident.

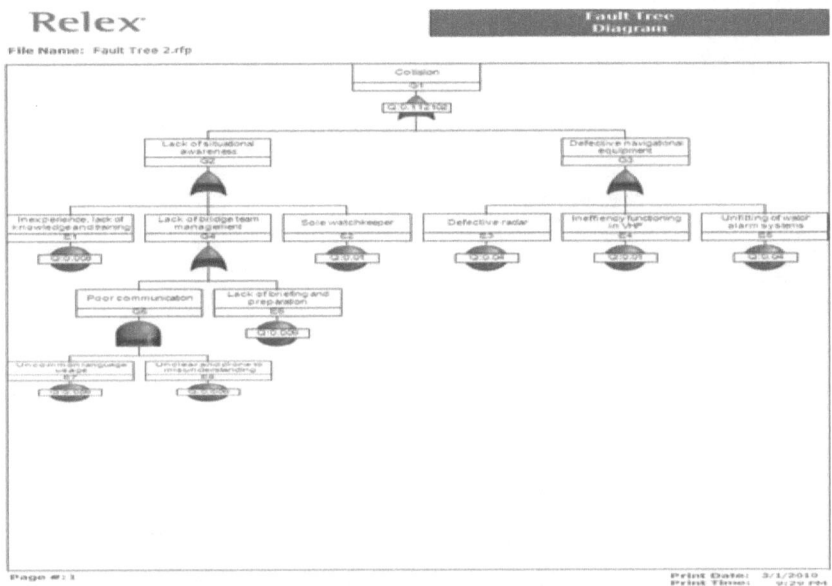

**Figure 4.84**: Fault tree diagram for the TE of collision from human error

Table 4.15 shows Fault tree calculation for collision TE.

**Table 4.15:** Fault tree calculation results for the top event of collision event

| Name | Description | Gate/Event Type | Input Type | FR/MTBF | Value 1 | Value 2 |
|------|-------------|-----------------|------------|---------|---------|---------|
| E1 | Inexperience, lack of knowledge and training | Basic Event | Constant Probability | | 0.008000 | |
| E2 | Sole watchkeeper | Basic Event | Constant Probability | | 0.010000 | |
| E3 | Defective radar | Basic Event | Constant Probability | | 0.040000 | |
| E4 | Inefficiency functioning in VHF | Basic Event | Constant Probability | | 0.010000 | |
| E5 | Unfitting of watch alarm systems | Basic Event | Constant Probability | | 0.040000 | |
| E6 | Lack of briefing and preparation | Basic Event | Constant Probability | | 0.009000 | |
| E7 | Uncommon language usage | Basic Event | Constant Probability | | 0.009000 | |
| E8 | Unclear and prone to misunderstanding | Basic Event | Constant Probability | | 0.009000 | |
| G1 | Collision | OR Gate | | | | |
| G2 | Lack of situational awareness | OR Gate | | | | |
| G3 | Defective navigational equipment | OR Gate | | | | |
| G4 | Lack of bridge team management | OR Gate | | | | |
| G5 | Poor communication | AND Gate | | | | |

Dr. Oladokun S. Olanrewaju
Ab Saman A. Kader

Numerical Calculation of ETA is:

Collision P(G1)is given by
$$= 1 - [1 - 0.0268][1 - 0.0876] = 0.1121$$
$$Q_2 = 1 - [1 - (0.1)(0.5)(0.2)(0.4)] = 0.004$$

Figure 4.85 shows an ETA analysis of HRA for collision event.

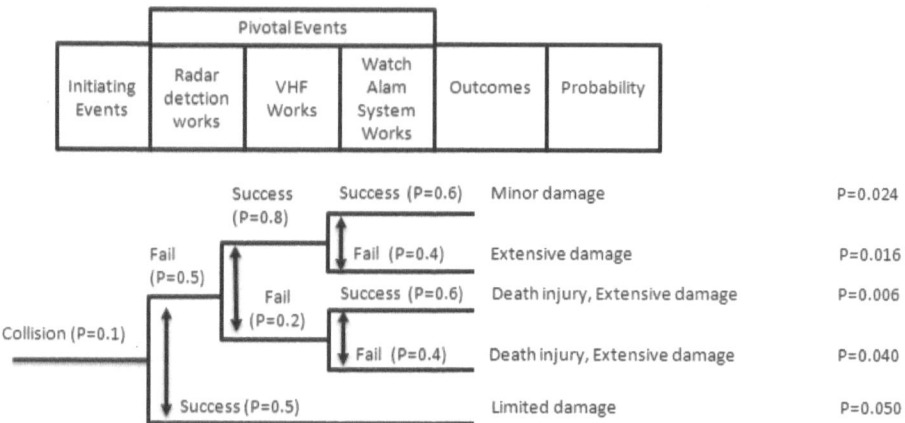

**Figure 4.85**: Event Tree Analysis of HRA for collision event

## 4.11.3 Criticality

Figure 4.86 shows risk criticality analysis vessel of 25, 000t is highly intolerable in the channel. Speed of 7 now is acceptable, but speed becomes critical 11 knot. Figure 4.86 shows that risk criticality analysis vessel of 25, 000T mass is highly intolerable in the Langat River. Speed of 7 now is acceptable, but speed becomes critical 11 knot. By studying the risk models associated with collision scenarios, four sub models in particular stands out where further risk reduction could be effective. Various channel components enter channel complexity components. These can be visibility weather, squat, bridge, river bent, human reliability. It is important to account for each of them in channel design work. Collision Risk index probability for visibility is obtained from: Fog Collision Risk Index (FCRI) = $(P_1 + VI_1 + P_2 + VI_2 + P_3.VI_3)$. Where: $P_k$ = Probability of collision per million encounters, $VI_k$ = Fraction of time that the visibility is in the range k, K = Visibility range: clear (>4km), Mist/Fog (200m to4km), Tick/dense (less than 200m).

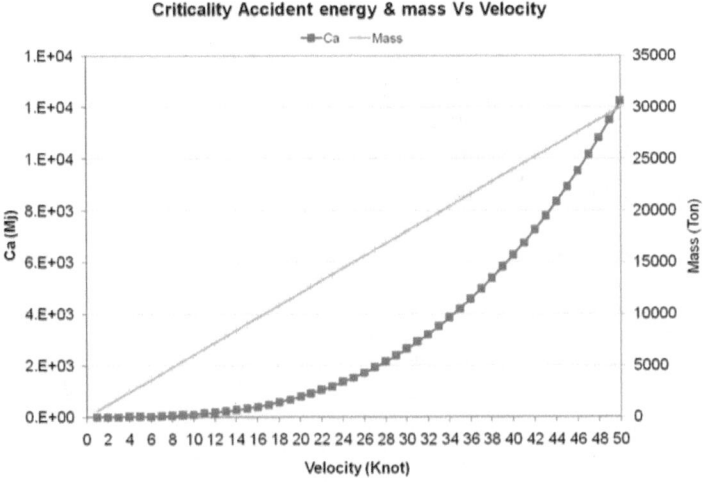

**Figure 4.86**: Criticality analysis

## 4.11.4 Visibility

Figure 4.87, a b, and c show visibility impact to collision, there is potential for higher risk as speed increase. Pc for appropriate visibility can be selected for lowest risk. There is potential for higher risk as speed increase. Pc for appropriate visibility can be selected for lowest risk. Figure 4.87a and 4.87b show visibility impact to collision, there is potential for higher risk as speed increase. Pc for appropriate visibility is selected for lowest risk. There is potential for higher risk as speed increase. Pc for appropriate visibility is selected for lowest risk.

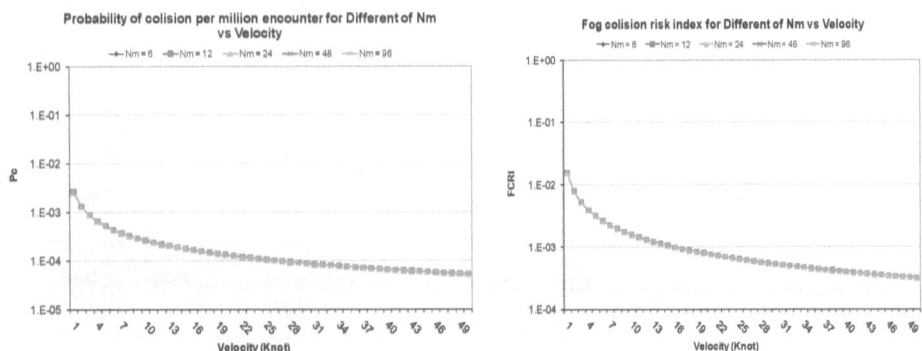

**Figure 4.87a**: Channel visibility analysis,
**Figure 4.88b**: Channel visibility Analysis for (FCRI)

## 4.11.5 Squat

Figure 4.88 shows squat analysis for Langat River. Figure 4.88 show squat allowances for ship speed of 3 knot, and beam of 21m, clearance for squat is—3. That gives ideas about requirement for dredging the channel, adjusting ship types and number.

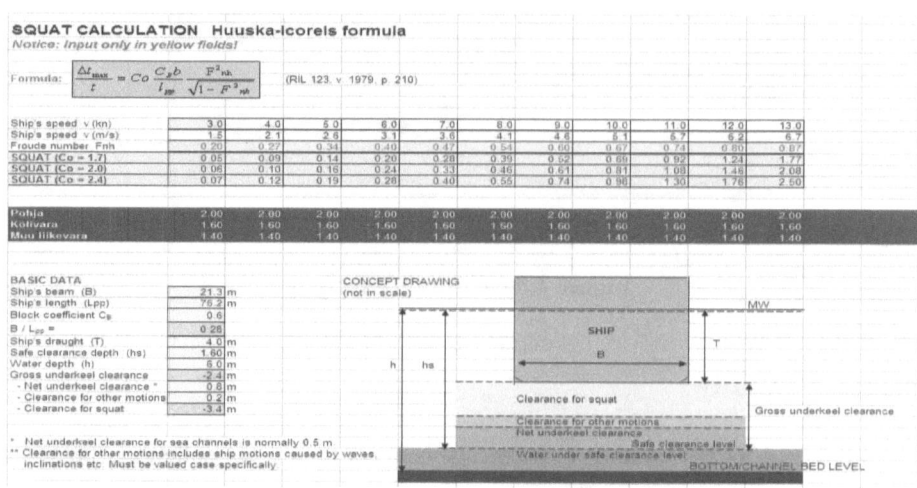

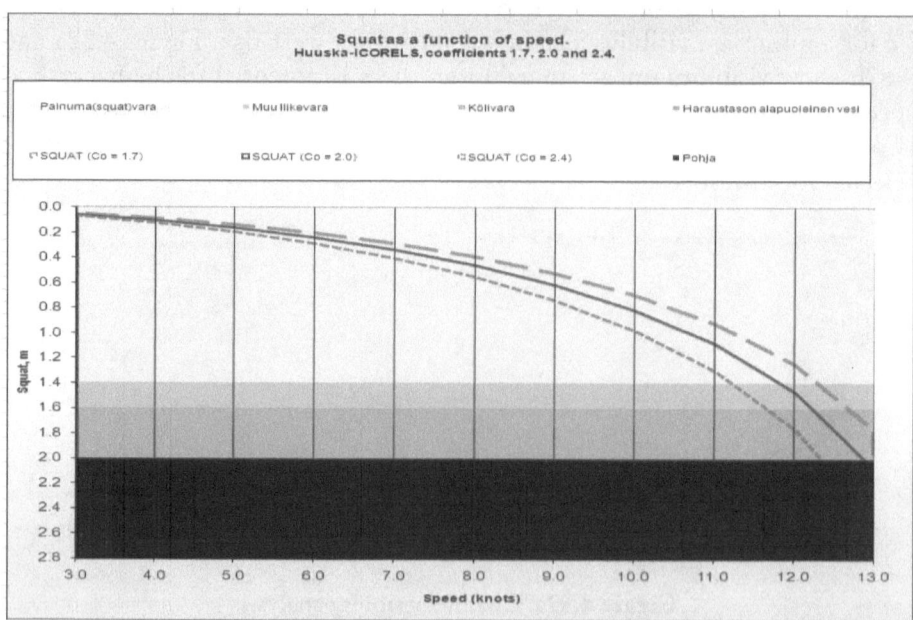

**Figure 4.88:** Squat analysis for Langat River

## 4.11.6 Result of Mooring and Berthing Analysis

Figure 4.89 show PIANC Mooring risk acceptability curve. Figure 4.89 shows PIANC acceptability requirement for berthing energy. where risk level are defined by Easy berthing (sheltered), difficult berthing, easy berthing exposed medium risk, Good berthing with exposed acceptable risk, Difficult berthing with exposed high risk . Berthing energy involve berthing velocity for the ships displacement range of 1000 tons to 10,000 t of water displacement is considered.

The highest speed value of the berthing velocity occur when the 1000t displacement ship berthing at the berthing facilities which is 0.6667 m/sec, the higher the value of water displacement of the ship, the lower the value of berthing velocity 0.653knot is considered best operating speed for vessel of 2000t and 0.5267knot is considered optimal speed for vessel of 5000 tons (Figure 4.90). Considering ships of 1000 to 10,000 tons displacement, the optimal berthing energy for the 2000 tons displacement ship berthing is 104.1 tm and for 5000 is 94.2t (Figure 4.91).

Environmental loading involve situation where the wind angle is taken from 30° to 175° with the average 30° as interval. The maximum lateral wind load occurs when the wind is extreme at 90° this has the value of 3109.75 pounds. The maximum longitudinal wind load occurs when the wind is extreme at 175° which has the value of 1500.52 pounds.

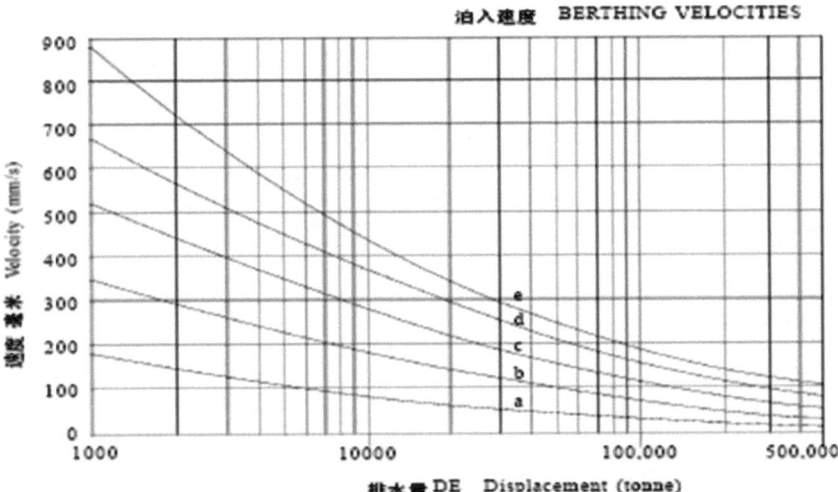

**Figure 4.89:** PIANC Standard for berthing velocity, a. Easy berthing (sheltered) b. difficult berthing (berthing) c. easy berthing (exposed), d. Good berthing (exposed), Difficult berthing (exposed)

Figure 4.90 shows berthing velocity, Figure 4.91 shows the berthing energy

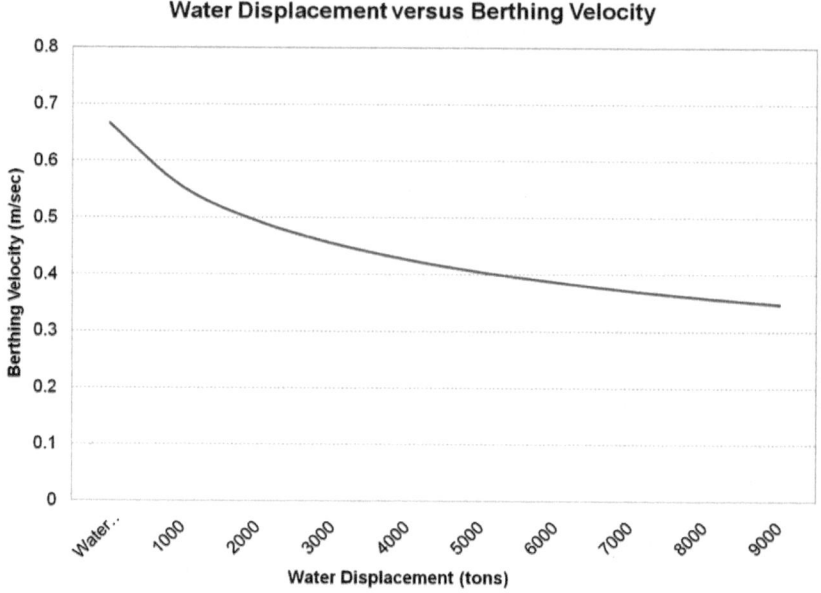

**Figure 4.90:** Berthing and water displacement

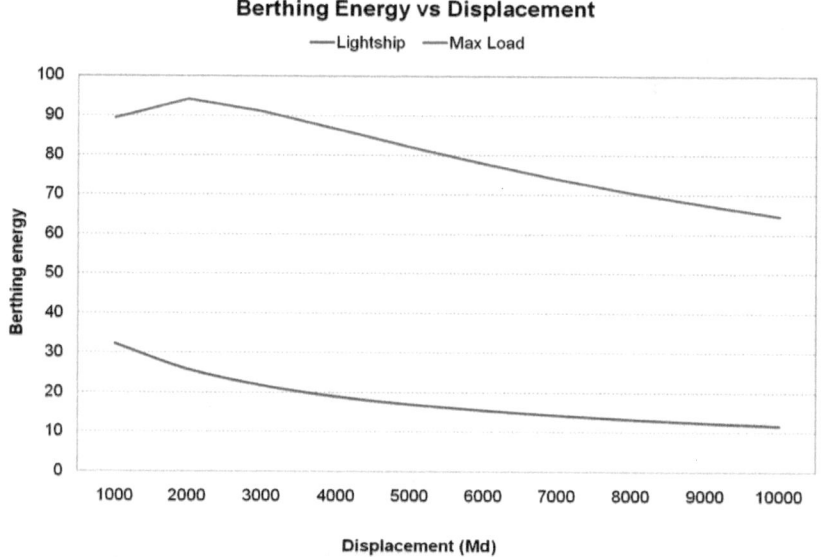

**Figure 4.91:** Berthing energy and displacement

Figure 4.92 shows mooring environmental loading. Figure 4.92a can be considerably acceptable as the amount may have little impact in the vessel dynamic positioning. The current angle is taken from 30° to 175° with the average 30° as interval. The maximum lateral current load occurs at the berthing maximum draft and current angle of 90° which has the value of 112.34 pounds. The maximum longitudinal current load occurs at berthing minimum draft and current angle of 145° which has the value of 12.21 pounds (Figure 4.92b).

TSS require consideration for symmetry of the channel, therefore analysis could considered reduction of parameters by half. This is determined by taking the standard deviation according to channel dimension.

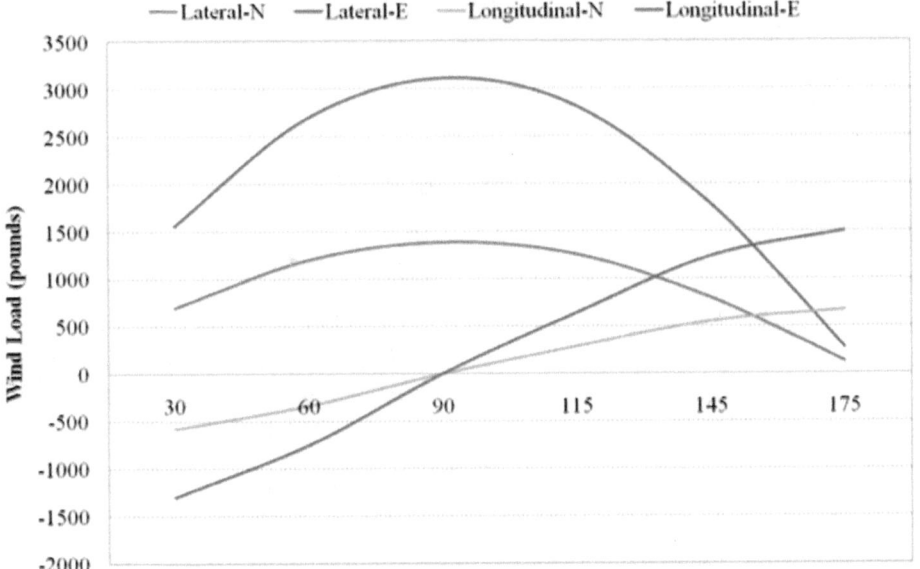

a.

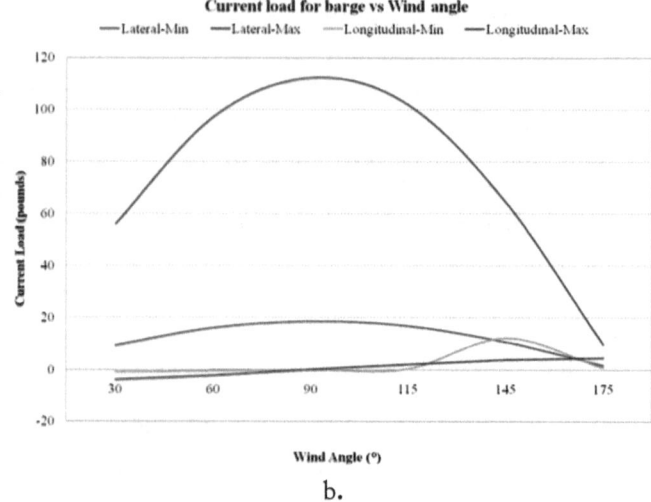

b.

**Figure 4.92:** a. Wind angle versus wind load, b. Current angle versus current load

## 4.11.7 Traffic Separation Scheme Analysis

Figure 4.93 a and b show TSS for different Collision Situation. Figure 4.93 show that TSS implementation, at various traffic scenarios, comparison of TSS shows that there is remnant risk for random situation is work for random and over taking situation and no TSS is required for crossing and head on collision. Channel can accommodate more number of vessels; acceptable number of ship is 7.

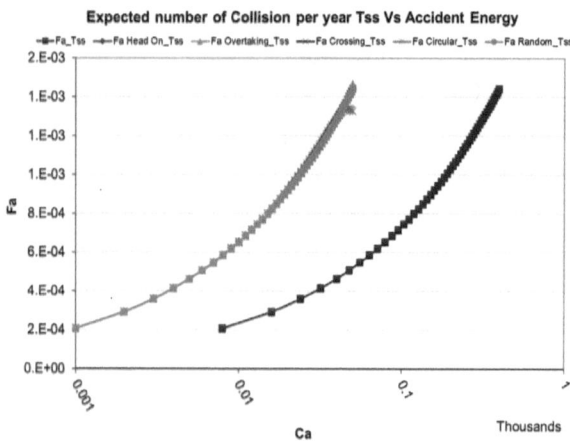

**Figure 4.93:** Accident frequency at changing number of vessel (TSS)

Figure 4.94.shows combined graph of speed with TSS and without TSS.

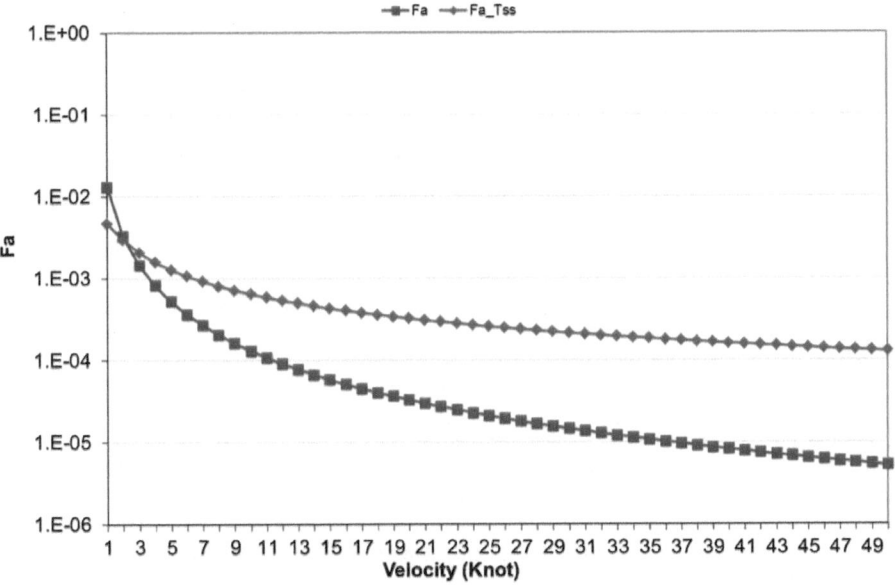

**Figure 4.94**: TSS and speed of ship, higher risk is observed for channel without TSS considering the same speed.

Figure 4.95 shows Combined Graph of Width with TSS and without TSS. Figure 4.94 describes combined graph of speed with TSS and without TSS. Channel can accommodate less speed with TSS due to channel width restriction. (Refer Figure 4.95). Larger width of channel is required for TSS. Figure 4.96 shows combine chart comparing effect of TSS and TSS on beam of ship. TSS depends on beam of ship, the higher the beam of ship the higher the risk for TSS. Figure 4.97 shows combined graph of changing number of ships with and without TSS. More number of ships can be accommodated for TSS (Refer to Figure 4.98), at increase number of ship, there is higher risk for without TSS. Figure 4.99 shows combined chart of variables for TSS and no TSS. In the generic model an as well as case study analysis, it is observed that TSS is not useful for inland waterways.

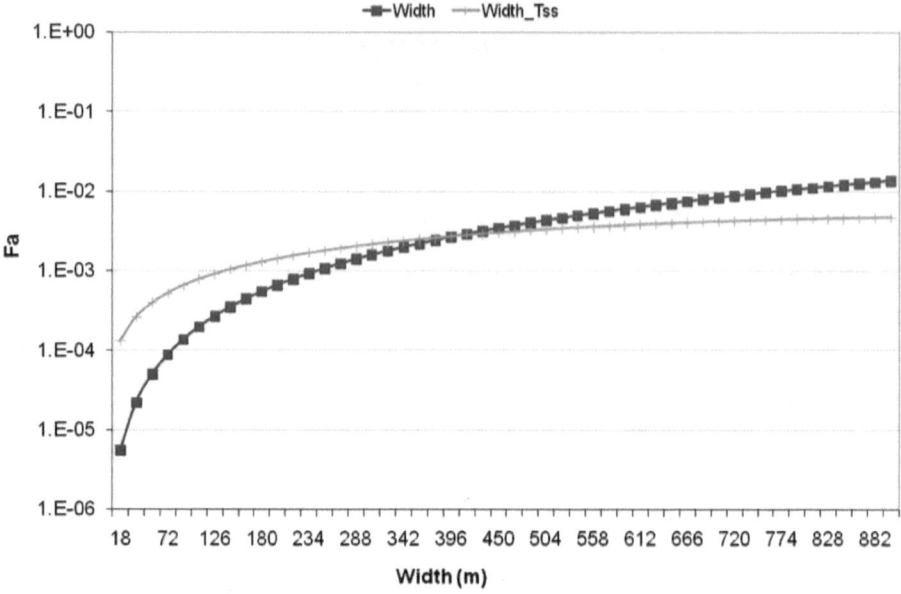

**Figure 4.95**: Width comparison, with TSS and without TSS

Figure 4.96 shows combine chart comparing effect of TSS and TSS on beam of ship.

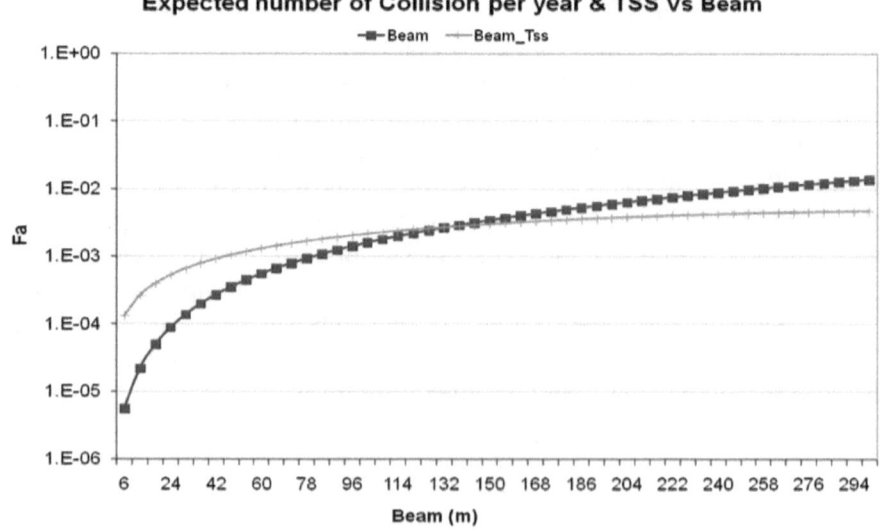

**Figure 4.96**: TSS and beam of ship

Figure 4.97 show combined graph of changing number of ships with and
without TSS

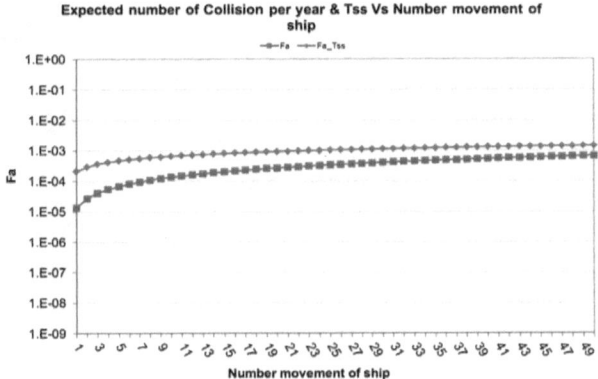

**Figure 4.97**: TSS and number of traffic density

Figure 4.98 shows combined chart of variables for TSS

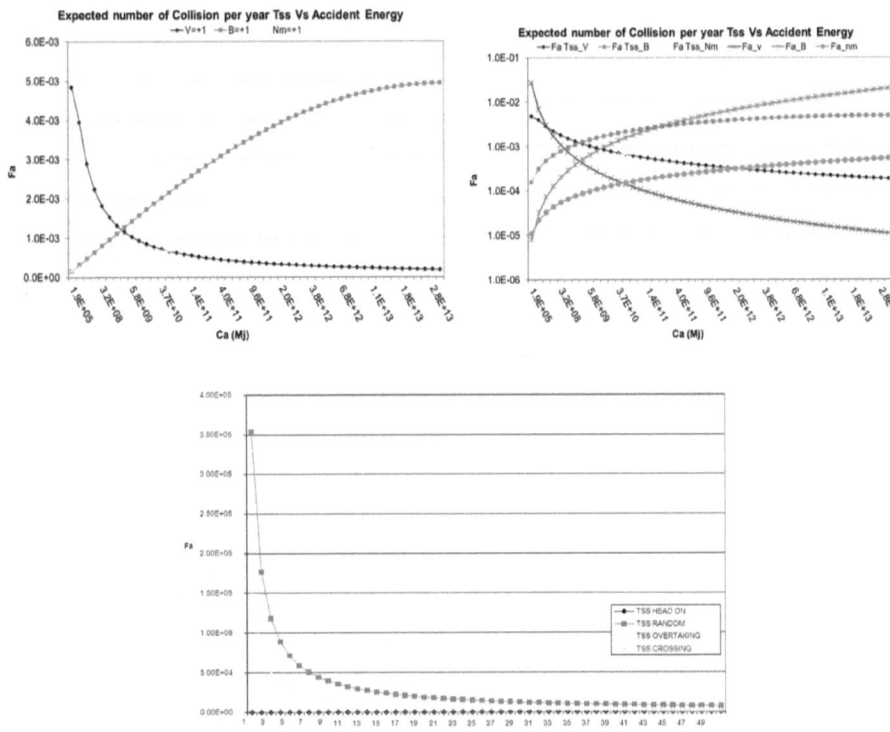

**Figure 4.98:** Interaction of parameters in TSS

# 5

---

# Conclusion

Inland water transport is precious natural resources considered one of the oldest economically and environmentally sustainable modes of transportation that involve vessels ranging from simple, non motorized boats to highly automated pushers, operating on waterways. Recent years have seen the drive toward the use of non renewable natural source replacing renewable sources which requires more energy and produces more waste. IWT is increasingly becoming important, and its implementation remains one of the biggest challenges for the 21st Century. The challenges include environmental sensitivity and need for proactive innovative behavior in design, construction maintainability and operations of system. This come with requirement to deal with inherent need and response require in the direction of harnessing nature energy reserve from use of sun, water, clean energy stored in earth crust and use of IWT. The closeness of Inland Water Transportation to large population make the need of risk based design an important requirement for IWT development.

Operation of IWT is critical, accident occurrence, damage and spill is prohibitive in coastal water due to high consequence and losses that comes with those events. This make it very imperative and necessary to design sustainable, efficient and reliable IWT that consider environmental, safety aspect of navigation channel, vessels and uncertainty development and implementation of proactive sustainable development philosophy. The SARM developed in this study provide answer to that need and future development of IWT.

Past engineering work on IWT has been dominated with reactive manner that focus on economic, technical and operational aspects of IWT. This study on safety risk and reliability model for IWT consider incorporated total risk systems engineering tools backed with hazard identification, scientific based qualitative and quantitative risk analysis that will secure alternative path to implementation philosophy of sustainability and reliability for use of IWT. This includes efficiency in carry out functions like shipping cargo containers through lock upgrades, replacement, intermodal, river information technology system, congestion tax break, and hybrid use of various identified alternatives. The holistic and system approached used in this study, make the SARM a good foundation for reliable development of IWT.

Risk and reliability based design and system integration are among requirement being promoted as integral part for sustainable and efficient using renewable natural resources like IWT. Well managed river basin can augment food water supplies, improve transportation, provide energy and develop the industry. The generic model for IWT is developed based on strategies for harbor, shipboard safety and environmental prevention and conservation thrust for treatment, elimination that involve Pollution Prevention or Pollution Control. These represent the backbone of the objective in achieving clean marine system design. Pollution prevention requires avoiding collision, as its occurrence result to release of environmentally harmful substances pollution control requires, preventing collision event through control of risk that leads to collision. Testing of the case study on the generic model, yield similar result that further strengthens validity of the model.

The research addresses risk contributing factors from maritime activities (Inland Water Transportation) that led to the development of a comprehensive probabilistic risk and reliability model for sustainable generic IWT. The use of PIANC waterway classification boundary work well for the generic model development, the generic model is implemented on Langat River. Previous studies on Langat River like many IWT under JPS master plan study for use of Langat water resources focused on economic, environmental data collection, and investigation of potential for use of Langat River for navigation. But no scientific based and system engineering risk work carried out in these studies has been used for investigation of risk and reliability in movement of vessel in the waterways. In order to extensively investigate the risk associated with IWT towards system reliability and life cycle accountability. The positive agreement of the trend of SARM deduced from scientific risk based design for IWT remain a good strength of the SARM model for its use for future development of accident free IWT.

The safety and environmental risk and reliability (SARM) model developed in this study involve risk assessment and analysis of collision accident scenario for prevention, control of collision, protection and conservation of the environmental. The model is developed identified risk required to avoid accident occurrence and consequence of collision of ships in waterways. The model address elements that can optimise design, existing practice, and innovative entity and facilitate decision support for policy accommodation for evolving coastal transportation regime.

IWT operation attracts high probability and consequence of accident which make reliability requirement for its design and operability for safety and environmental protection very necessary. Risk and reliability based design entails the systematic integration of risk analysis in the design process. It target risk prevention, reduction as part of design objective. Waterways accident falls under scenario of collision and grounding, fire and explosion, flooding. Collision carries the highest percentage and more frequent. Collision carries the largest risk scenarios in maritime industry. Among the cause of cause are propulsion function, navigation failure and human error, which is agreement with the model developed.

The model was built with design from first principle. Design from first principle deterministic method has been utilized by incorporating natural laws in the risk process. The study embarked on the task to capture the IWT physical system. The study of physical system provide opportunity to capture the pressure, response and action associated with IWT development, subsequently examine, evaluated the current situation hazard and changes require for the development of IWT. Analysis from system potential risk and comparative advantage lead to deduction of high level goal based objective is required to improve system strength and reliability. The cause of system failure is much due to lack of sensitivity analysis on the system environmental components. Qualitative analysis has been undertaken. HAZID tools were used based on level of degradation potential that enter IWT risk regime to identified hazard event occurrence, detection and severity. System (global and local) database are assessed and evaluated with standards, community satisfaction, system capability that leads to high level objective and parameters consideration. PIANC recommendation for IWT development and IMO rule making risk model framework adopted from UK HSE and offshore industry to mitigate accident risk in IWT are used as strong foundation for the development of SARM generic model for IWT.

The predicted result is subject to reliability test. Reliability validation of SARM provide valuable and effective decision support tool for application of automated system based safety analysis that can be used to facilitate incorporation of reliable safety control as part of the iterative design processes objective for new and innovative marine system designs including IWT. The model result represents a tool for safer, more reliable marine system and products including IWTS. Total risk approach and cost quantification of benefit are derived through inductive scientific and stochastic approach that further give ground for use of the model for decision support for IWT development. In SARM development, frequency and consequence are quantitatively analysed to generate ALARP graph that accommodate total risk components variables simulation required for relevant decision support. Accident collision per year has been determined. The ALARP principle used in the model represents quality of a SARM model total risk characteristics.

Damage in term of volume of collapse and length of collapse is estimated from the model collision energy analysis. Also, Interpolation from energy analysis are to other consequence in term of oil spill and carbon dioxide emission has been performed to extract other uncertain consequence risk information within the system. Risk measurement for reliability is obtained by determining risk control option. This is followed by determination of cost effectiveness analysis as well as sustainability balance between environmental, economic and safety. Implications of quantifying cost and benefit attached to waterways as part of decision support is analysed. Model testing and validation by statistical method is undertaken to quantify confidence, reliability and credibility in the analytic predictive model SARM model developed.

The model is further validated by case study using Langat River. Langat River is considered for the model case study validation and feasibility of its implementation because of the fact regarding the potential for its use for transportation and also because of emphasis made by LUAS, that collision is their major problem on Langat River. Putting a large body of river like Langat to work through a reliability based design that mitigate generic safety and environmental risk of collision prevent of collision will support substantial decision for its use and other waterways.

Technical and environmental capability is incorporated based on hazard, degradation or damage, and the role of new innovative eco technological to improve existing methods. Possible energy consequence releases from collision event in IWTS are translated from energy of hit collision analysis performed in the SARM model. CEA is utilised to analysed risk and benefit components of the system. The model also examined components of channel complexity and uncertainty to determine remnant risk in the system. CEA leading to required risk control measure and option for the system is determined. Reliability based validation is performed to test the accuracy of the model. Optimization of predictive accuracy of the model can be done through additional experiments, information, and experience under expert rating and community participation program.

The study incorporated uncertainty and channel complexity factors. Deduce risk and risk control option. RCO and CEA deduced also take into consideration release of C0x and oil spill. The result of the model generated is interesting; the model used PIANC classification of waterways data and IMO and HSE guideline. The ALARP curves for generic and case study as well as testing on statistical software, are iterated several time and yielded repeatable trend. Overall the ALARP risks are in low risk region.

The need to reduce time and costs associated with large-scale physical system drive the reliance on analytical modeling and simulation upward. The tools developed in a reliable SARM model developed are encouraged to be continuously developed for improvement of system requirement and observation under monitoring regime. Stochastic quantification of the confidence and prediction accuracy of the model provides the decision support information necessary for adoption of the SARM model for development complex system like IWT risk aversion of associated infrequent high consequence decisions support.

The SARM model output can be used as part of algorithms needed to quantify the uncertainties associated with the IWTS. Future research and development recommended for improvement of the model include continuous work on improvised methods and tools for performing uncertainty quantification. Use of further peer-reviewed for validation hierarchy metric, and reliability performance, utilization of model for other waterways as required according to environmental matching. Conversion of the model to simulation software, especially virtual reality that will further present behavior of the SARM for efficient analysis of IWTS risk and reliability analysis is recommended.

# 6

# References

A.S.A. Kader, Ab. Kader. (1997). Cost Modeling for Inland Waterway Transportation System. PhD Thesis. Liverpool. pp 186.

A.S.A. Kader, Adi Maimun, and A. Haris. (2004). River Transportation In Malaysia. A Case Study of Klang River, For Push – Tow System". Kuching , Sarawak

Afifi and Ramzan. (1999). Application of New Technology in Inland Water Transport. National Seminar on inland Water Transport. Kuching. Sarawak.

Akita Y., N. Ando, Y. Fujita, and K. Kitamura. (1972). Studies on Collision-Protective Structures In Nuclear Powered Ships. Nucl Eng Design 19. pp. 365–401.

Asian Development Bank (ADB). (1994). Klang River Basin Integrated Flood Mitigation Project. Final Report. Vol. 3. Malaysia.

Ayyub, B.M., Beach, J.E., S. Sarkani. Assakkaf. I.A. (2002). Risk Analysis and Management for Marine System. Naval Engineers Journal. Vol. 114, No. 2, pp. 181-206.

Abbas.B.M. (1983). River basin development. Tycooly. Dublin.

Bottelberghs. P.H. (1995). QRA in the Netherlands. Conference on Safety Cases, IBC/DNV, London Report of marine accident. Access on 14, March 2009.

Broils. (1967.). J.U. New European norms for Size of Waterway Urgently Needed. Hinterland ports. Rotterdam Europort Delata.

Brown A, Amrozowicz M. (1996).Tanker environmental risks—putting the pieces together. International Conference on Design and Methodologies for Collision and Grounding Protection of Ships. San Francisco, USA.

Butts, Thomas A. and Dana B. Shackleford. (1992). Pacts of Commercial Navigation on Water Quality in the Illinois River Channel. ISWS RR-122. USA

Brebbia C.A & J. Olivella. Gallor. W. (2000). The Safety Of Ship Movement In Port Water Area. Maritime Engineering and Port. WIT press. pp. 80-89.

Cahill, R.A. Collisions and Their Causes. (1983).. London. England. Fairplay Publications.

Carr, J. H. (1952) Wave Protection Aspects of Harbor Design. Report E-11. Hydrodynamics Laboratory, California Institute of Technology.

Catherine Hetherington, Rhona Flin, and Kathryn Mearns. (2006). Safety in Shipping: The Human Element. Journal of Safety Research. 37, 401–411.

Clay, Grady (ed.). (1979). Water & the Landscape. London. pp 193.

Coleman, H.W., W.G. Steele, Jr. (1989). Experimentation and Uncertainty Analysis for Engineers. John Wiley & Sons.

CBO. (1982) U.S. Congress. Energy Use in Freight Transportation. Washington. DC. pp. 10.

Cooke, R.M. (1997). Uncertainty Modeling: Examples and Issues. Safety Science. Elesevier. pp. 49-60.

Stapersma D. (1996). Diesel Engines: A fundamental Approach to Performance Analysis, Turbo Charging, Combustion Emissions and Heat Transfer. Part A: Performance analysis. United Kingdom.

Stamatis. D.H. (2004). Failure Mode Effect Analysis from theory to execution. ASQ. pp. 112-169

David Evans and Associates. Inc. (2001). Corridor Program Draft Fish and Aquatic

Habitat. Expertise Report. Bellevue. Washington. I-405.

Vose David (1996). Risk Analysis. A Quantitative Guide. John Wiley & Sons, INC. Canada pp. 67-87.

JPS (2000). Modeling and Data Integration in the Study of Sediment. Kuala Lumpur. Malaysia.

DON. (1984). Harbors Design Manual. NAVFAC DM-26.1. Alexandria, VA.USA.

Devanney, J.W., III (1967). Marine Decisions Under Uncertainty. MIT Press. Cambridge, MA.

Det Norske Veritas (DnV). (2000). Thematic Network for Safety Assessment of Waterborne Transportation. Norway.

DnV . (2003). Formal Safety Assessment of cruise navigation. Report No. 2003-0277. Norway. "FSA – Large Passenger Ships – Navigational

Safety". IMO NAV 51/10. Norway, International Maritime
Organization. Norway.

DnV. (2001). Marine Risk Assessment. Her majesty stationary office United
Kingdom.

Dharam Ghai and Jessica M. Vivian (eds.). (1992).Grassroots Environmental
Action: People's Participation in Sustainable Development. Routledge.
London.

Eastman. S.E. (1980). Fuel Efficiency in Freight Transportation. The
American Waterway Operators Inc. Arlington. VA. pp.7.

Eftratios Nikolaidis. (2005). Probabilistic analysis of dynamic systems.
Engineering design reliability. CRC press. pp. 17-1.

Emi, H. et al. (1997). An Approach to Safety Assessment for Ship
Structures. ETA analysis of engine room fire, Proceedings of ESREL '97.
International Conference on Safety and Reliability. Vol. 2.

Eryuzlu, N.E., Cao. Y.L., and D'Agnolo. F (1994). Under keel Requirements
for Large Vessels in Shallow Waterways. PIANC Proceedings 28th
International Congress. Section II. Subject 2. Belgium.

EC (2006). Trans European Transport Network Report. Indicative
Programme. Brussels.

EC. (2006). European Transport Network Report. Indicative Programme.
Brussels.

Evans SM, Leksono T, McKinnell PD. (1995). Tributyltin Pollution: a
Diminishing

Problem Following Legislation Limiting the Use of TBT Based Antifouling
Paints. Mar. Pollution Bull. 30 (1). 1995. pp.14–21.

Eryuzlu, N.E., Cao. Y.L., and D'Agnolo, F. (1994).Under keel Requirements
for Large Vessels in Shallow Waterways. PIANC Proceedings 28th
International Congress. Section II. Subject 2.

Caho. F. S. B. (1996) Shipping and the Environment. Is Compromising
inevitable. IMarEST London.

Bercha F.G. and M. Cerovšek, and Wesley Abel. (2004). Reliability
Assessment of Arctic EER Systems. International Association for
Hydraulic Engineering and Research.

Fotland, H. (2004). Human Error: A Fragile Chain of Contributing
Elements. The International Maritime Human Element Bulletin. No. 3,
pp. 2–3. Published by the Nautical Institute. London, U.K.

Funtowicz, S. O., and Ravetz. J. R. (1992). Three Types Of Risk Assessment
And The Emergence Of Post-Normal Science. p. 251-274. Westport, CT.

GESAMP. (1997). Impact of Oil and Related Chemicals and Wastes on
Marine Environment. GESAMP reports and studies No50 joint group

of expert of marine pollution. Available from: http://www.gesamp.imo. org/no65/. Access on June 1997

Guedes Soares, C., A. P. (2001). Teixeira. Risk Assessment in Maritime Transportation. Reliability Engineering and System Safety. pp. 299-309.

Harwich, J. B. Harbors. (1992). Navigational Channels, Estuaries. Environmental Effects. Handbook of Coastal and Ocean Engineering, Vol. 3. Gulf Publishing Company. Houston, TX. 1992. pp. 1340.

Henningsen, R.F. (2000). Study of Greenhouse Gas Emissions from Ships. Final report to the International Maritime Organization. MARINTEK. Trondheim, Norway

Hollnagel, E. (2005). CREAM: Cognitive Reliability an Error Analysis Method.

HSE (2001) Reducing Risk, Protecting People: HSE's decision-making process. Health & Safety Commission

F. S. B. Caho. (1996). Shipping and the Environment. is Compromising inevitable. IMarEST London.

F.G. Bercha and M. Cerovšek, and Wesley Abel. (2004). Reliability Assessment of Arctic EER Systems. International Association for Hydraulic Engineering and Research.

Fotland, H. (2004). Human Error: A Fragile Chain of Contributing Elements. The International Maritime Human Element Bulletin. No. 3, pp. 2–3. Published by the Nautical Institute, 202 Lambeth Road, London, U.K.

Fowler, T.G. et al (2000). Modeling Ship Transportation Risk. Risk Analysis, 20(2): p. 225-244.

Funtowicz, S. O., and J. R. Ravetz. (1992). Three types of risk assessment and the emergence of post-normal science. p. 251-274. Westport, CT.

Fujii Y. (1974). Some factors affecting the frequency of accidents in marine traffic. II—the probability of stranding and III—the effect of darkness on the probability of collision and stranding. J Navigation 27 pp. 21–43.

Fujii Y. (1982). Recent Trends in Traffic Accidents in Japanese Waters. Journal of Navigation. Vol35 (1). pp. 88—102.

GESAMP. Impact of Oil and Related Chemicals and Wastes on Marine Environment.

GESAMP reports and studies No50 joint group of expert of marine pollution. Available at: http://www.gesamp.imo.org/no65/. Access on June 1997

Guedes Soares. C., A. P. Teixeira. (2001). Risk Assessment in Maritime Transportation. Reliability Engineering and System Safety. pp. 299-309.

Harwich, J. B. (1992). Harbors, Navigational Channels, Estuaries. Environmental Effects. Handbook of Coastal and Ocean Engineering, Vol. 3. Gulf Publishing Company. Houston, TX. pp. 1340.

Henningsen, R.F. (2000). Study of Greenhouse Gas Emissions from Ships. Final report to the International Maritime Organization. MARINTEK, Trondheim, Norway.

Hollnagel, E. (2005). CREAM: Cognitive Reliability an Error Analysis Method. Retrieved on 14 August 2009, at *http://www.ida.liu.se/~eriho/ CREAM_M.htm*.

Hufschmidt, M.M. and K.G. Tejwani. (1993). Integrated Water Resources Management: Meeting the Sustainability Challenge. UNESCO. Humid Tropics Programme (IHP) Series No. 5. UNESCO. Paris. France.

Illinois State Water Survey. (1992). Department of Energy and Natural Resources.

Impacts of Commercial Navigation on Water Quality in the Illinois River Channel. Champaign. IL.

ILO. (1976). Guide to Safety and Health in Dock Work. Amended in 1979.

IMO (2003). MSC Circular 1091. Issues to be Considered when Introducing New on Board Ships.

IMO. (2009). Annex VI of MARPOL 73/78. Regulations for the prevention of air pollution from ships and NOx technical code. Publication IMO-664E, London UK.

IMO. (ISM Code). (1994). Adopted by IMO, Resolution A.741 (18). UK

IMO. (2002). Convention on the International Regulations for Preventing Collisions at Sea 1972. Consolidated edition 2002.

IMO. (1993). International Code for Ships Carrying Liquefied Gases in Bulk (IGC Code). 1994/1996 Amendments to the IGC (replaced the Gas Carrier Code)

IMO. (1991). International Convention for Prevention of Pollution from Ships (MARPOL) 1973/78. Consolidated 1991 and later amendments. Including 1997 Annex VI. Prevention of Air Pollution. UK

IMO. (1978). International Convention on Standards of Training. Certification and Watch keeping (STCW) for Seafarers. UK

IMO. (1996). STCW '95. International Maritime Organization. London, UK. ISBN: 92—801-1412-3.

IAGLP. ( 1972. ) Report of the Engineering Committee. The St. Lawrence Seaway: The Quiet. Efficient Marine Highway. Toronto. pp.5.

IMO. (2004)International Convention for the Safety of Life at Sea (SOLAS). (2004). Consolidated Edition. London.

IMO. (2006). Amendments to the Guidelines for Formal Safety Assessment (FSA) for Use in the IMO Rule Making Process. MSC – MEPC.2/Circ 5 (MSC/Circ.1023 – MEPC/Circ.392).

IMO. (1994). International Convention for the Safety of Life at Sea (SOLAS). 2006 Amendments. Chapter II-2. Part D. Reg. 13 Means of Escape.

IMO. (2001). Guidance on Fatigue Mitigation and Management. MSC/CIRC. 1014. London

Jenseb, Soren. (1996). Report of New Chemical Hazard. New Scientist. Principle of environmental science and technology. 32. pp. 612.

John X. Wang. (2000). What Every Engineer Should Know about Risk Engineering and Management. Markel Deker Inc. Switzerland. pp. 112-128.

Jonh R. Dudley. (1994). Towards Safer Seas & Cleaner Seas. London. pp. 139-151.

Kite Powell, H. L., D. Jin, N. M. Patrikalis, J. Jebsen, V. Papakonstantinou. (1996). Formulation of a Model for Ship Transit Risk". MIT Sea Grant Technical Report. Cambridge, MA, 1996. pp 96-19.

Klanac, A., Ehlers, S., Tabri, K., Rudan, S., and Broekhuijsen, J. (2005). Qualitative design assessment of crashworthy structures, International Congress of International Maritime Association of the Mediterranean (IMAM). Lisboa, Portugal, 26–30

Kitamura O. (1996). Comparative study on collision resistance of side structure. International Conference on Design and Methodologies for Collision and Grounding Protection of Ships. San Francisco.

Laurel Gascho, Henrike Peichert, and Sarah Renner. (2006). Malaysia Referral & Comparative Experiences of Inland Waterway Transportation System. Environment and Poverty Networks.

Lee, S.G., et al, H. (2001). Crashworthy structural design of bowstructure, 15th TEAM, Jochiwon. Korea.

Lempert, R. J., S. W. Popper and S. C. Bankes. (2003). Shaping the Next One Hundred Years: New Methods for Quantitative Long-Term Policy Analysis. RAND: Santa Monica, CA. pp. 187.

Logan, R.L., and C.K. Nitta. (2002). Verification & Validation (V&V) Guidelines and Quantitative Reliability at Confidence (QRC): Basis for an Investment Strategy. Lawrence Livermore National Laboratory.

Lord Donaldson. (1994). Safer Ship Cleaner Ocean. Report of Lord Donalson enquiry into the prevention of pollution from merchant ships. London.

Lind N.C.(1996). Safety Principle and Safety Culture. 3rd International Summit on Safety at Sea. Norway.

Luna B. Leopold. (1994). A View of the River. Harvard University Press. ISBN. A Non Technical Primer on the Geomorphology and Hydraulics of Water.

Moderras M.. (1993). What Every Engineer Should Know about Reliability and Risk Analysis. Markel Deker Inc. Switzerland. pp. 299-314.

MOSTE. (2000). Malaysia initial national communication. United Nation Framework convention on climate change. Kuala Lumpur

Ohtsu. Masaki (2007). Possible change of ship design including engine room about emission of NOx and SOx. Mitsui Engineering & Shipping Co Ltd. Japan

McGee, S.P., Troesch, A., and Vlahopulos. (1999). N. Damage Length Predictor for High Speed Craft. Marine Technology. Vol. 36(4). pp. 201-214.

Millward, A. (1990). A Preliminary Design Method for the Prediction of Squat in Shallow Water. Marine Technology 27(1):10-19.

Mohd. Zamani b.Ahmad (1999). Multimodalism and the Role of Inland Water System as and Infrastructure. National Seminar on Inland Water Transport. Kuching. Sarawak.

Macduff, T. (1974). The probability of vessel collisions. Ocean Industry 9(9): 144–148.

Minorsky, V.U. (1959). An Analysis Of Ship Collisions With Reference To Nuclear Power Plants. Ship Res 3 2 (1959), pp. 1–4.

Murphy, D.M. & M.E. Paté-Cornell. (1996). The SAM Framework. A Systems Analysis Approach to Modeling the Effects of Management on Human Behavior in Risk Analysis. pp. 501-515.

NWF. (1983). U.S. Waterways Productivity. A Private and Public Partnership. Huntsville. AL. pp. 165-167.

National Research Council (2000). Risk Management in the Marine Transportation System. Conference Proceedings. Washington, DC: National Academy Press.

Nathwani, J. S., Lind, N. C., Pandey, M. D. (1997) Affordable safety by choice: the life quality method. University of Waterloo, Institute for Risk Research. Canada.

NMD. (2009). Concerning the medical examination of employees on Ships. 200909387/hri/ucl. Norway

University of Michigan. (2006). Operator Fatigue. Department of Naval Architecture and Marine Engineering. Retrieved on 12 May 2008, from http://www.ardujenski.com/files/documents/FatigueDesign.pdf

OECD. (1990) European Conference of Ministers of Transport. Policy and the Environment. Paris, France. pp. 99.

Osterreichische Wasserstrassen. (2007). Inland Environmental Performance. RINA. pp. 49.

Parry, G. (1996). The Characterization of Uncertainty in Probabilistic Risk Assessments of Complex Systems. Reliability Engineering and System Safety". 54:2-3. pp. 119-126.

Cornell Pate M.E. (1996). Uncertainties in Risk Analysis: 6 Levels of Treatment. Reliability Engineering and System Safety. pp. 95-111.

Peddicord, R. K., Chase, T., Dillon, T., McGrath, J., Munns, W. R., Van De Gucthe, K., Van Der Schalie. Ecological risk assessment of contaminated sediments. Society of Environmental Toxicolog. W. Workgroup Summary Report on Navigational Dredging. Brussels.

Peterson, P.J. (1997). Indicators of Sustainable Development in Industrializing Countries. Vol II. Lestari Monograph. UKM. Bangi.

Pedersen P.T. (1993). Ship Impacts: Bow Collisions. Int J Impact Eng 13pp. 163-187.

PIANC (2002). Working Group Envicom 16: Dredging and Port Construction. Brussels.

PIANC. (2002). Guidelines for the Design of Fender Systems. Brussels.

PIANC (1992). Capacity Of Ship Maneuvering Simulation Models For Approach. Brussels.

PIANC(1997). Guidelines for the Design of Armored Slopes under Open Piled Quay Walls. Supplement to Bulletin No. 96. Permanent International Association of Navigation Congresses. Brussels. Belgium.

PIANC (2007). Incom Inland Waterways Commission, Annual Report. Brussel, Belgium.

Pittock Wratt B. D. (2001). Australia and New Zealand. In Climate Change: Impacts, Adaptations, and Vulnerability. Contribution of Working Group II to the Thirds Assessment Report of the International Panel on Climate Change. Chapter 12.

Psaraftis, H.N. (2002) Maritime Safety: To Be or Not to Be Proactive. WMU Journal of Maritime Affairs. 1. pp. 3-16.

Rackwitz, R. (2002). How Safe Is Safe Enough? An approach by Optimisation and Life Quality Index. Proceeding of ASTRANET Conference.

Raiffa, H. (1969). Preference for multi-attributed alternatives. RM-5868-DOT/RC. The RAND Corporation Santa Monica, CA.

Rausand, M (2005). Preliminary Hazard Analysis. Department of Production and Quality Engineering, Norwegian University of Science and Technology. Retrieved on 11 April 2008, from *http://www.ntnu.no/ross/srt/slides/pha.pdf*

Regu k. Draisamy. (1999). Lightweght Composite Sandwich Structures. Workshop lightweight Composite Structure. Bangalore. Retrieved on 14 August 2009, from *http://imv.au.dk/~pba/Preprints/HumFact*

Roach, P.J. (1998). Verification and Validation in Computational Science and Engineering. Hermosa Publishers. Albuquerque. NM.

Roeleven D. M., Kok H. L. , Stipdonk de Vries. W. A. (1995). Inland Waterway Transport: Modeling the Probabilities of Accidents. Safety Science. 1995. pp 191-202.

Schubel, J. R. (1972). Distribution and Transportation of Suspended Sediment at a Station in the Chesapeake Bay. Environmental framework of Coastal Plain Estuarine. Geological Society of American Memoir, 133. pp.151 – 167.

Sirkar J, Ameer P, Brown A, Goss P, Michel K, Nicastro F, Willis W. A. (1997) Framework For Assessing The Environmental Performance of Tankers In Accidental Groundings and Collisions. SNAME annual meeting.

Skjong, R. (2002). Risk Acceptance Criteria: Current Proposals. IMO Position. Proceedings.

Skjong, R., Vanem, E. & Endersen. (2005). Risk Evaluation Criteria. SAFEDOR Deliverable D.4.5.2.

Sulaiman O. (2006). Sustainable Maintenance of Navigation Channel. M. Eng. Thesis. Skudai, Johor.

Sylvia Earle. (1995). Sea Change. A Message of the Oceans.

Texas Transportation Institute. (2007). Center for Ports and Waterways. A Modal Comparison of Domestic Freight Transportation Effects on the General Public. MARPOL 73/78, consolidated edition 1997. IMO 520E. November.

Thomas Koester. (2003). Human Factors and Everyday Routine in the Maritime Work Domain. Danish Maritime Institute Hjortekærsvej. DK-2800 Lyngby. Denmark.

Trbojevic, V.M., Carr, B.J. (2000). Risk Based Methodology For Safety Improvements In Ports. Journal of Hazardous Materials 200. 71 (1–3). pp. 467–480.

TTI. (2007). A Modal Comparison of Domestic Freight Transportation Effects on The General Public. Center for Ports and Waterways. Texas

USACE. (1980). Institute for Water Resources. Water Resources Support Center, National Waterways Study. Analysis of Environmental Aspects of Waterway Navigation. Review Draft. Fort Beloit. pp. 227. VA.

USCG. (1999). Regulatory Assessment: Use of Tugs to Protect Oil Spill in the Puget Sound Area.

U.S. Department of Commerce. (1974). Maritime Administration. A Modal Economic and Safety Analysis of the Transportation of Hazardous Materials In Bulk. Cambridge. MA. pp.9.

UNEP, (1992). Environmental Data Report. Basil Blackwell. Oxford. pp 143.

UNEP (1998). World's Surface Water: Precipitation, Evaporation and Runoff. In UNEP/GRID—Arendal Maps and Graphics Library.

Van der Leeden Frits. (1990). The Water Encyclopedia. 2nd ed. Chelsea, MI: Lewis Publishers.

Volpe national Transportation Center. (1997). Scoping Risk Assessment, Protection Against Oil Spill in the Marine Waters of Northwest Washington State.

Wallace Blischke. R. (2000). Reliability Modeling, Prediction, and Optimization. pp. 119-123.

Wang G, Chen Y, Zhang H, Shin Y. (2000). Residual strength of damaged ship hull. Ship Structure Symposium on Ship Structures for the New Millennium: Supporting Quality in Shipbuilding, Arlington. VA.

World Resources Institute (WRI). (2001). United Nations Development Programme and World Bank. World Resources. People and Ecosystems: The Fraying Web of Life. World Resources Institute. Washington DC.